On Being the Right Size
and other essays

On Being the Right Size and other essays

J. B. S. HALDANE

EDITED BY
JOHN MAYNARD SMITH

Oxford New York
OXFORD UNIVERSITY PRESS

Oxford University Press, Walton Street, Oxford OX2 6DP

London New York Toronto
Delhi Bombay Calcutta Madras Karachi
Kuala Lumpur Singapore Hong Kong Tokyo
Nairobi Dar es Salaam Cape Town
Melbourne Auckland

and associated companies in
Beirut Berlin Ibadan Mexico City Nicosia

Oxford is a trade mark of Oxford University Press

For details of previous publication see Acknowledgements
This selection first published as an Oxford University Press paperback 1985
Reprinted 1985

British Library Cataloguing in Publication Data
Haldane, J. B. S.
On being the right size and other essays.
1. Science
I. Title II. Smith, John Maynard
500 Q158.5
ISBN 0-19-286045-3

Library of Congress Cataloging in Publication Data
Haldane, J. B. S. (John Burdon Sanderson), 1892-1964
On being the right size and other essays.
Includes index.
1. Biology—Addresses, essays, lectures.
2.—Science—Addresses, essays, lectures.
I. Smith, John Maynard, 1920- II. Title.
QH311.H318 1985 574 84-16578
ISBN 0-19-286045-3 (pbk.)

Printed in Great Britain by
The Guernsey Press Company Ltd
Guernsey, Channel Islands

ACKNOWLEDGEMENTS

THE publisher would like to thank the previous publishers of these essays for permission to include them in the present volume. Permission has been sought from the original publisher where known, but where details of original publication have been difficult to ascertain it has been sought from the publishers of the collections in which the essays in question appeared. The following list gives details of the relevant prior publication of all the essays.

Essays marked with an asterisk were previously collected in *Possible Worlds and other essays* (Chatto & Windus, London, 1927), those marked with a dagger in *The Inequality of Man and other essays* (Chatto & Windus, London, 1932): in neither collection are precise details of original publication given. Those marked with ‡ appeared in *Science and Everyday Life* (Lawrence & Wishart, London, 1939), and were originally published in the *Daily Worker*; and those marked with § appeared in *A Banned Broadcast and other essays* (Chatto & Windus, London, 1946).

* 'On Being the Right Size'
* 'William Bateson'
* 'The Future of Biology'
* 'When I am Dead': originally one of a series on this topic, most of which supported a different thesis
* 'Science and Theology as Art-Forms'
* 'The Last Judgement'

† 'Is History a Fraud?'
† 'God-Makers'
† 'The Origin of Life'
'The Biology of Inequality': the first chapter of *Heredity and Politics* (Allen & Unwin, London, 1938)
‡ 'Beyond Darwin'
‡ 'The Strange Case of Rahman Bey'

‡ 'Blood and Iron'
‡ 'Germ-Killers'
‡ 'Pain-Killers'
§ 'How to write a Popular Scientific Article': *Journal of the Association of Scientific Workers*
§ 'What "Hot" Means': *Daily Worker*
'Cats': *Daily Worker*; reprinted in *Science Advances* (Allen & Unwin, London, 1947)
'Some Reflections on Non-Violence': *Mankind* 1.4 (1956), published in India

The publisher has also made every effort to trace the heir to the copyrights in the work published by the author in his lifetime, with a view to obtaining the appropriate permission, but without success at the time of going to press. Any information leading to an identification will be gratefully received.

CONTENTS

INTRODUCTION

JOHN MAYNARD SMITH

I FIRST came across J. B. S. Haldane's essays when, as a schoolboy at Eton, I found that he was the person my schoolmasters most hated. Feeling that anyone they hated that much could not be all bad, I went to seek his books in the school library. It is, I suppose, to my masters' credit that I found *Possible Worlds and other essays.* The impact of those essays was such that, fifteen years later, when I decided to leave engineering and train as a biologist, I entered University College London, where Haldane was at that time a professor, and became his student and later his colleague.

I can recall two things from that first reading of Haldane's essays. The first, gained from the essay 'On Being the Right Size', which is reproduced here, was the realization that there is a bridge between mathematics and natural history; Haldane's own major contribution to science was in helping to build that bridge. My second memory is of the intelligent barnacles in the title essay, 'Possible Worlds', which distinguish between 'real' objects, which they can reach with their arms, and 'visions', which they can see but not reach, and of the philosophers belonging to the species *Lepas sapiens* who have devised a means of predicting whether (and if so, for whom) a vision will turn into a real object, by performing abstruse calculations on the directions in which the visions are perceived by different barnacles. I have omitted 'Possible Worlds' from this collection, partly because it is rather long, and partly because I hope to provoke readers to seek out that collection for themselves.

As a scientist, Haldane will be remembered for his contribution to the theory of evolution. Today, Darwin's theory of natural selection and Mendel's theory of genetics are so intimately joined together in 'neo-Darwinism' that it is hard to imagine that, after the rediscovery of Mendel's laws in 1900, the two theories were seen as rivals. Haldane, together with R. A. Fisher and Sewell Wright, showed that they were compatible, and developed the theory of population genetics which still underpins all serious thinking about evolution. However, although it is not hard to identify Haldane's major contribution to science, he is in other respects somewhat difficult to classify. A liberal individualist, he was best known as a leading communist and contributor of a weekly article to the *Daily Worker*. A double first in classics and mathematics at Oxford, he made his name in biochemistry and genetics. A captain in the Black Watch who admitted to rather enjoying the First World War, he spent the end part of his life in India writing in defence of non-violence.

These contradictions in his character and opinions only make him more interesting to read. He was supreme as a popularizer of science, because he saw connections that others missed. The present selection ranges from articles intended for a scientific readership to others written for a daily newspaper, the only constraint being that they should all be comprehensible without any special technical knowledge. I have tried to cover the full range of his interests. Perhaps inevitably, essays on religious topics may be over-represented. However, I do not think that this amounts to a distortion; Haldane was at the same time fascinated and repelled by religious ideas. It is characteristic of him that, when he settled in India after his retirement from University College London, he should have made such a serious effort to understand Hinduism.

It is also characteristic of him that he used popular articles to propose original ideas; some examples are included here. 'The Origin of Life' is perhaps the most important. The ideas presented here contributed significantly to the fact that the topic is now one for experimental study and not for philosophical speculation. The essay 'When I am Dead' makes the claim (which I think he would have rejected later in life) that if mental processes are physically determined, there is no reason to suppose that their conclusions are correct, a claim which has since been elaborated by Karl Popper. I discuss three other original ideas in the Appendix, 'Adumbrations', in which I give brief quotations from Haldane's writings, together with an account of what has happened to the ideas since.

Haldane's views changed during his lifetime, although I think that his views on practical and political matters changed more than his basic philosophy. I have arranged the essays, at least approximately, in chronological order, so that changes in opinion are more readily recognized. However, I must emphasize that my choice has been aimed at illustrating the characteristic cast of Haldane's mind rather than at recording his political history. The main feature of that history, of course, is that he joined the communist party in the late thirties, and left it in the fifties. His political views can be seen in the present volume in the set of articles from *Science and Everyday Life*, *A Banned Broadcast and other essays*, and *Science Advances*, all but one of which originally appeared in the *Daily Worker*. They differ stylistically from the other essays in being aimed at a readership with less formal education, and they contain a number of explicitly political remarks, but I cannot see that they differ significantly in their attitude to science. I do not think that anyone today is writing scientific articles for the daily Press which approach these either in scientific content or entertainment value. To illustrate their scientific content, the article 'Beyond Darwin' contains a particularly clear account of Darwin's idea about the relationship between sexual dimorphism and polygyny, which has since become a popular topic of research. I was amused to find that the same article contains a false argument (about species which destroy their food supply and starve to death). I learnt the fallacy of such 'group-selectionist' arguments from Haldane himself when I was his student.

In general, I do not think that Haldane's conversion to Marxism had much effect on the way he saw biology, or on his views about how science ought to be done. However, it does seem to have forced him to rethink his approach to human genetics. Although highly critical of eugenic policies, as proposed in Britain and applied in the United States, his early essays reveal a somewhat hereditarian approach. I have here reprinted the first chapter from *Heredity and Politics*, published in 1938, which presents a more sophisticated discussion of the nature–nurture problem; it is a treatment which many contemporary writers on the subject have still not understood.

After some hesitation, I have decided not to add footnotes explaining particular references in the essays. Some of these will be unfamiliar to the reader: indeed, some are so to me. The general sense of Haldane's argument is so clear, however, that footnotes could do nothing to make it easier to follow.

Rereading Haldane's essays has for me been an extraordinary pleasure, which I hope you will share.

ON BEING THE RIGHT SIZE

THE most obvious differences between different animals are differences of size, but for some reason the zoologists have paid singularly little attention to them. In a large textbook of zoology before me I find no indication that the eagle is larger than the sparrow, or the hippopotamus bigger than the hare, though some grudging admissions are made in the case of the mouse and the whale. But yet it is easy to show that a hare could not be as large as a hippopotamus, or a whale as small as a herring. For every type of animal there is a most convenient size, and a large change in size inevitably carries with it a change of form.

Let us take the most obvious of possible cases, and consider a giant man sixty feet high—about the height of Giant Pope and Giant Pagan in the illustrated *Pilgrim's Progress* of my childhood. These monsters were not only ten times as high as Christian, but ten times as wide and ten times as thick, so that their total weight was a thousand times his, or about eighty to ninety tons. Unfortunately the cross-sections of their bones were only a hundred times those of Christian, so that every square inch of giant bone had to support ten times the weight borne by a square inch of human bone. As the human thigh-bone breaks under about ten times the human weight, Pope and Pagan would have broken their thighs every time they took a step. This was doubtless why they were sitting down in the picture I remember. But it lessens one's respect for Christian and Jack the Giant Killer.

To turn to zoology, suppose that a gazelle, a graceful little creature with long thin legs, is to become large; it will break its bones unless it does one of two things. It may make its legs short and thick, like the rhinoceros, so that every pound of weight has still about the same area of bone to support it. Or it can compress its body and stretch out its legs obliquely to gain stability, like the giraffe. I mention these two beasts because they happen to belong to the same order as the gazelle, and both are quite successful mechanically, being remarkably fast runners.

Gravity, a mere nuisance to Christian, was a terror to Pope, Pagan, and Despair. To the mouse and any smaller animal it presents practically no dangers. You can drop a mouse down a thousand-yard mine shaft; and, on arriving at the bottom, it gets a slight shock and walks away. A rat is killed, a man is broken, a horse splashes. For the resistance presented to movement by the air is proportional to the surface of the moving object. Divide an animal's length, breadth, and height each by ten; its weight is reduced to a thousandth, but its surface only to a hundredth. So the resistance to falling in the case of the small animal is relatively ten times greater than the driving force.

An insect, therefore, is not afraid of gravity; it can fall without danger, and can cling to the ceiling with remarkably little trouble. It can go in for elegant and fantastic forms of support like that of the daddy-long-legs. But there is a force which is as formidable to an insect as gravitation to a mammal. This is surface tension. A man coming out of a bath carries with him a film of water of about one-fiftieth of an inch in thickness. This weighs roughly a pound. A wet mouse has to carry about its own weight of water. A wet fly has to lift many times its own weight and, as everyone knows, a fly once wetted by water or any other liquid is in a very serious position indeed. An insect going for a drink is in as great danger as a man leaning out over a precipice in search of food. If it once falls into the grip of the surface tension of the water—that is to say, gets wet—it is likely to remain so until it drowns. A few insects, such as water-beetles, contrive to be unwettable, the majority keep well away from their drink by means of a long proboscis.

Of course tall land animals have other difficulties. They have to pump their blood to greater heights than a man and, therefore, require a larger blood pressure and tougher blood-vessels. A great many men die from burst arteries, especially in the brain, and this danger is presumably still greater for an elephant or a giraffe. But animals of all kinds find difficulties in size for the following reason. A typical small animal, say a microscopic worm or rotifer, has a smooth skin through which all the oxygen it requires can soak in, a straight gut with sufficient surface to absorb its food, and a simple kidney. Increase its dimensions tenfold in every direction, and its weight is increased a thousand times, so that if it is to use its muscles as efficiently as its miniature counterpart, it will need a thousand times as much food and oxygen per day and will excrete a thousand times as much of waste products.

Now if its shape is unaltered its surface will be increased only a hundredfold, and ten times as much oxygen must enter per minute through each square millimetre of skin, ten times as much food through each square millimetre of intestine. When a limit is reached to their absorptive powers their surface has to be increased by some special device. For example, a part of the skin may be drawn out into tufts to make gills or pushed in to make lungs, thus increasing the oxygen-absorbing surface in proportion to the animal's bulk. A man, for example, has a hundred square yards of lung. Similarly, the gut, instead of being smooth and straight, becomes coiled and develops a velvety surface, and other organs increase in complication. The higher animals are not larger than the lower because they are more complicated. They are more complicated because they are larger. Just the same is true of plants. The simplest plants, such as the green algae growing in stagnant water or on the bark of trees, are mere round cells. The higher plants increase their surface by putting out leaves and roots. Comparative anatomy is largely the story of the struggle to increase surface in proportion to volume.

Some of the methods of increasing the surface are useful up to a point, but not capable of a very wide adaptation. For example, while vertebrates carry the oxygen from the gills or lungs all over the body in the blood, insects take air directly to every part of their body by tiny blind tubes called tracheae which open to the surface at many different points. Now, although by their breathing movements they can renew the air in the outer part of the tracheal system, the oxygen has to penetrate the finer branches by means of diffusion. Gases can diffuse easily through very small distances, not many times larger than the average length travelled by a gas molecule between collisions with other molecules. But when such vast journeys—from the point of view of a molecule—as a quarter of an inch have to be made, the process becomes slow. So the portions of an insect's body more than a quarter of an inch from the air would always be short of oxygen. In consequence hardly any insects are much more than half an inch thick. Land crabs are built on the same general plan as insects, but are much clumsier. Yet like ourselves they carry oxygen around in their blood, and are therefore able to grow far larger than any insects. If the insects had hit on a plan for driving air through their tissues instead of letting it soak in, they might well have become as large as lobsters, though other considerations would have prevented them from becoming as large as man.

Exactly the same difficulties attach to flying. It is an elementary principle of aeronautics that the minimum speed needed to keep an aeroplane of a given shape in the air varies as the square root of its length. If its linear dimensions are increased four times, it must fly twice as fast. Now the power needed for the minimum speed increases more rapidly than the weight of the machine. So the larger aeroplane, which weighs 64 times as much as the smaller, needs 128 times its horsepower to keep up. Applying the same principles to the birds, we find that the limit to their size is soon reached. An angel whose muscles developed no more power weight for weight than those of an eagle or a pigeon would require a breast projecting for about four feet to house the muscles engaged in working its wings, while to economize in weight, its legs would have to be reduced to mere stilts. Actually a large bird such as an eagle or kite does not keep in the air mainly by moving its wings. It is generally to be seen soaring, that is to say balanced on a rising column of air. And even soaring becomes more and more difficult with increasing size. Were this not the case eagles might be as large as tigers and as formidable to man as hostile aeroplanes.

But it is time that we passed to some of the advantages of size. One of the most obvious is that it enables one to keep warm. All warm-blooded animals at rest lose the same amount of heat from a unit area of skin, for which purpose they need a food-supply proportional to their surface and not to their weight. Five thousand mice weigh as much as a man. Their combined surface and food or oxygen consumption are about seventeen times a man's. In fact a mouse eats about one-quarter its own weight of food every day, which is mainly used in keeping it warm. For the same reason small animals cannot live in cold countries. In the arctic regions there are no reptiles or amphibians, and no small mammals. The smallest mammal in Spitzbergen is the fox. The small birds fly away in the winter, while the insects die, though their eggs can survive six months or more of frost. The most successful mammals are bears, seals, and walruses.

Similarly, the eye is a rather inefficient organ until it

reaches a large size. The back of the human eye on which an image of the outside world is thrown, and which corresponds to the film of a camera, is composed of a mosaic of 'rods and cones' whose diameter is little more than a length of an average light wave. Each eye has about half a million, and for two objects to be distinguishable their images must fall on separate rods or cones. It is obvious that with fewer but larger rods and cones we should see less distinctly. If they were twice as broad two points would have to be twice as far apart before we could distinguish them at a given distance. But if their size were diminished and their number increased we should see no better. For it is impossible to form a definite image smaller than a wavelength of light. Hence a mouse's eye is not a small-scale model of a human eye. Its rods and cones are not much smaller than ours, and therefore there are far fewer of them. A mouse could not distinguish one human face from another six feet away. In order that they should be of any use at all the eyes of small animals have to be much larger in proportion to their bodies than our own. Large animals on the other hand only require relatively small eyes, and those of the whale and elephant are little larger than our own.

For rather more recondite reasons the same general principle holds true of the brain. If we compare the brain-weights of a set of very similar animals such as the cat, cheetah, leopard, and tiger, we find that as we quadruple the body-weight the brain-weight is only doubled. The larger animal with proportionately larger bones can economize on brain, eyes, and certain other organs.

Such are a very few of the considerations which show that for every type of animal there is an optimum size. Yet although Galileo demonstrated the contrary more than three hundred years ago, people still believe that if a flea were as large as a man it could jump a thousand feet into the air. As a matter of fact the height to which an animal can jump is more nearly independent of its size than proportional to it. A flea can jump about two feet, a man about five. To jump a given height, if we neglect the resistance of the air, requires an expenditure of energy proportional to the jumper's weight. But if the jumping muscles form a constant fraction of the animal's body, the energy developed per ounce of muscle is independent of the size, provided it can be developed quickly enough in the small animal. As a matter of fact an insect's muscles, although they can contract more quickly than our own, appear to be less efficient; as otherwise a flea or grasshopper could rise six feet into the air.

And just as there is a best size for every animal, so the same is true for every human institution. In the Greek type of democracy all the citizens could listen to a series of orators and vote directly on questions of legislation. Hence their philosophers held that a small city was the largest possible democratic state. The English invention of representative government made a democratic nation possible, and the possibility was first realized in the United States, and later elsewhere. With the development of broadcasting it has once more become possible for every citizen to listen to the political views of representative orators, and the future may perhaps see the return of the national state to the Greek form of democracy. Even the referendum has been made possible only by the institution of daily newspapers.

To the biologist the problem of socialism appears largely as a problem of size. The extreme socialists desire to run every nation as a single business concern. I do not suppose that Henry Ford would find much difficulty in running Andorra or Luxemburg on a socialistic basis. He has already more men on his payroll than their population. It is conceivable that a syndicate of Fords, if we could find them, would make Belgium Ltd. or Denmark Inc. pay their way. But while nationalization of certain industries is an obvious possibility in the largest of states, I find it no easier to picture a completely socialized British Empire or United States than an elephant turning somersaults or a hippopotamus jumping a hedge.

WILLIAM BATESON

IF the Proceedings of the Brunn[1] Natural History Society had been a little rarer I suppose that Bateson would now be lying in Westminster Abbey. For we have only to read between the lines of the first report to the Evolution Committee of the Royal Society by himself and Miss E. R. Saunders, published in 1902, to realize that when Mendel's paper in the Brunn Society's journal was discovered in 1900, Bateson had already hit upon the atomic theory of heredity, which goes by the name of Mendelism. It was characteristic of him that no hint of this fact is to be found in his published work. His classical exposition of the subject is entitled *Mendel's Principles of Heredity*. Copernicus, if he admitted Aristarchus' priority, did not write on 'Aristarchus' principles of astronomy'. But Mendel's and Bateson's discovery was as fundamental as that of Copernicus, and of much greater practical importance.

And yet Bateson was not of a retiring disposition. The early days of Mendelism were marked by extremely violent controversy on both sides—I can remember the time when Mendelism was considered grossly heretical at Oxford—a controversy in which Bateson played a notable part. And his public attacks on the Darwinian theory were so phrased as inevitably to lead to the most heated argument, and even to the extraordinary misrepresentation that he disbelieved in evolution. So far was this from being the case that if he had died thirty years ago he would be remembered mainly for his work on Balanoglossus, a worm-like marine creature which he showed to constitute a link between vertebrates and invertebrates.

[1] Brno since the war.

It was the wide and, as he felt, uncritical acceptance of the theory of evolution by natural selection which led him to expose its weak points. But it was eminently characteristic of him that he took up a not altogether dissimilar attitude to his own work. He had many disciples, but was never himself of their number. The characters which are inherited according to Mendel's laws are so numerous and important, and their possible combinations so enormous, that a lesser man would inevitably have devoted the remainder of his days to following out the detailed application of those laws. Bateson did so up to about 1912, but the last years of his life were largely given over to the investigation of exceptions to them; and we owe to him more than to any other one man the demonstration not only that they are valid over a vast range of material, but that they occasionally break down. His last published work deals with these exceptions, and their importance is exaggerated rather than minimized.

His mental processes were well illustrated by his attitude to the work of Morgan and his school in New York, who have shown that the Mendelian factors are carried in or by the chromosomes which can be seen in a dividing nucleus. For eight years Bateson attacked this theory with the utmost vigour; not because he considered it inherently improbable, but because he believed that it went beyond the evidence, and because the natural bent of his mind and his profound knowledge of the history of science led him to doubt the validity of long chains of reasoning, however convincing. When, however, the possibility of ocular demonstration arose, he went over to America, and returned a convert, though with certain reservations which I believe that the future will largely justify. It is the fact that he had retained his mental elasticity until the time of his death that makes that death so grievous a loss to biology.

Yet I can well believe that those who knew him but slightly carried away a different impression. He never attempted to conceal his contempt for second-rate work or second-rate thought, and pursued the truth with no more regard for other people's opinions than for his own. He started his career as a morphologist, and his outlook was always morphological. He was therefore sometimes unduly sceptical of reasoning from a non-morphological standpoint. But I never had an argument with him—and I had many—without the absolute conviction that he would no more hesitate to admit himself in the wrong if I could convince him, than to tell me that I was talking nonsense if, as was more usual, I failed to do so.

His scientific views inevitably led him to doubt the possibility of far-reaching improvements in human life by alteration of the environment. He was inclined to the belief that the best elements in the human race were being weeded out; and the mutual destruction of them which went on during the war confirmed him in it. But he regarded most if not all of the attempts to apply science to this problem by creating an art of eugenics as premature in view of our profound ignorance of human heredity, and resolutely refused to associate himself with eugenical organizations. From the pessimism which such views inevitably engendered he found a refuge not only in science but in art; and his exquisite sense of form drew him to the art of the far East, of which he was a well-known connoisseur.

If Bateson had merely demonstrated the truth and importance of Mendelian heredity the world would be his debtor. For in its essential manifestations it is so simple that I have known a child of fourteen apply it with complete success to practical breeding; and yet it furnishes the only clue that we have at present to innumerable problems concerning the nature of the cell, the course of evolution, the determination of sex, and even the origin of certain human races.

But Bateson did much more than that. He has probably

prevented Mendelism from becoming a dogma. For example he held that it would not, as some at least of his disciples believe, explain evolution. It is normal for a discoverer to be obsessed by the importance of his own discoveries, and it is a thoroughly excusable weakness. There are times in the history of thought when an idea must be born, and if it is a great idea it may be expected to overwhelm and obsess the man who gave it birth. He either becomes its slave, or preserves a certain independence only by continuing to hold views incompatible with it at the expense of dividing his mind into watertight compartments. William Bateson escaped these fates because he was greater than any of his ideas.

THE FUTURE OF BIOLOGY

IN forecasting the future of scientific research there is one quite general law to be noted. The unexpected always happens. So one can be quite sure that the future will make any detailed predictions look rather silly. Yet an actual research worker can perhaps see a little further than the most intelligent onlooker. Even so, it may seem presumptuous for any one man, especially one who is almost completely ignorant of botany, to attempt to cover, however inadequately, the whole field of biological investigation.

Every science begins with the observation of striking events like thunderstorms or fevers, and soon establishes rough connections between them and other events, such as hot weather or infection. The next stage is a stage of exact observation and measurement, and it is often very difficult to know what we should measure in order best to explain the events we are investigating. In the case of both thunderstorms and fever the clue came from measuring the lengths of mercury columns in glass tubes, but what prophet could have predicted this? Then comes a stage of innumerable graphs and tables of figures, the despair of the student, the laughing-stock of the man in the street. And out of this intellectual mess there suddenly crystallizes a new and easily grasped idea, the idea of a cyclone or an electron, a bacillus or an antitoxin, and everybody wonders why it had not been thought of before.

At present much of biology is in the stage of measuring and waiting for the idea. One man is measuring the lengths of the feelers of 2,000 beetles; another the amount of cholesterol in 100 samples of human blood; each in the hope, but not the certainty, that his series of numbers will lead him to some definite law. Another is designing a large and complicated apparatus to measure the electrical currents produced by a single nerve fibre when excited, and does not even look beyond the stage of the column of figures. If I were writing this article for biologists it would be largely a review of present and future methods; to a wider public I shall try to point out some of the results now emerging, and their possible application.

Let us begin with what used to be called natural history; the study of the behaviour of animals and plants in their wild or normal condition. Apart from animal psychology this has split up into two sciences, ecology and animal sociology. Extraordinary progress has recently been made in the latter. Wheeler of Harvard has made it very probable that the behaviour of social insects such as ants, instead of being based on a complicated series of special instincts, rests largely on an economic foundation not so very unlike our own. The ant that brings back a seed to the nest gets paid for it by a sweet juice secreted by those that stayed at home. Others, again, have been tackling the problem of how much one bee can tell another, and how it does it. Tomorrow it looks as if we should be overhearing the conversation of bees, and the day after tomorrow joining in it. We may be able to tell our bees that there is a tin of treacle for them if they will fertilize those apple trees five minutes' fly to the south-east; Mr Johnson's tree over the wall can wait! To do this we should presumably need a model bee to make the right movements, and perhaps the right noise and smell. It would probably not be a paying proposition, but there is no reason to regard it as an impossible one. Even now, if we take a piece of wasps' comb and hum the right note, the grubs put out their heads; if we then stroke them with a very fine brush they will give us a drop of sweet liquid just as they do to their nurses. Why should we wait to see if there are 'men' on Mars when we have on our own planet highly social and perhaps fairly intelligent beings with a means of communication? Talking with bees will be a tough job, but easier than a voyage to another planet.

In ecology, where we deal with animal and plant communities which consist of many different species, each eaten by others from inside and outside, each living in amity with some of its neighbours, in competition with others, we are at present often lost in detail. But we are constantly finding that some hitherto unexpected but often easily modifiable factor, such as the acidity of the soil or the presence of some single parasite on an important species, will make a whole new fauna and flora appear, say an oak forest with wild pigs instead of a pine forest with ants.

We apply these principles in agriculture by using chemical manures and insects parasitic on those that attack our crops. But as we find the key chemical or key organism in a given association, we may be able vastly to increase the utility to man of forests, lakes, and even the sea. Besides this, however, one gets the very strong impression that from the quantitative study of animal and plant associations some laws of a very unsuspected and fundamental character are emerging; laws of which much that we know of human history and economics only constitutes special and rather complicated cases. When we can see human history and sociology against a background of such simpler phenomena, it is hard to doubt that we shall understand ourselves and one another more clearly.

In the domain of classificatory zoology our ideal is to establish a family tree of plants and animals: to be able to say definitely, let us say, that the latest common ancestor of both man and dog was a certain definite type of animal living, for example, in what is now the North Atlantic 51,400,000 years ago, under the shade of the latest common ancestor of the palm and beech trees, while the last common ancestor of the dog and bear lived only 5,200,000 years back. We are still thousands of years from this ideal, but we are now attacking the problem of relationships between living forms by a number of new methods, especially chemical methods. For example, we find that man agrees with the chimpanzee and other tailless apes, and differs from the tailed monkeys, in being unable to oxidize uric acid to allantoin in his tissues, as well as in many anatomical characters. This merely confirms our view that these apes are man's nearest relations. But the same kind of method will be applied to solving problems of relationship in which the anatomical evidence is less clear; for example, what group of four-footed animals is most nearly related to the whale. Animals have a chemical as well as a physical anatomy, and it will have to be taken into account in their classification.

But the most important evidence about evolution is coming from the study of genetics. We take any animal or plant, and with sufficient time and money at our disposal should be able to answer the following questions (though if it is a slow-breeding animal like a cow it is more likely that our great-great-grandchildren will have to wait for the answer):

1. What inheritable variations or mutations arise in it and how are they inherited?

2. Why do they arise?

3. Do they show any sign of being mainly in any one direction, or of advantage to their possessor?

4. Would natural selection acting on such, if any, as are advantageous, account for evolution at a reasonable speed, and for the kind of differences which are found between species (for example, that which causes sterility in hybrids)?

The first question can often be answered, the second rarely. Occasionally we can provoke mutations, as with radium or X-rays. There is no indubitable evidence that they ever arise in children in sympathy with bodily changes in their parents (the alleged transmission of acquired characters), and plenty of well-established cases where they do not. Now, we know how the genes, or units which determine heredity, are arranged in the nucleus of the cell, and also about how big they are. If we magnified a hen's egg to the size of the world (which would make atoms rather larger than eggs and electrons barely visible) we could still get a gene into a room and probably on to a small table. But such magnification being impossible, the question how to interfere with a single gene without interfering with the others becomes serious, and some men have already spent their lives vainly on it; many more will. The two most hopeful methods seem to be to find chemical substances which will attack one gene and not another; and to focus ultraviolet rays on a fraction of a chromosome, the microscopic constituent of the nucleus in which the genes are packed. One can focus ultraviolet rays far more exactly than ordinary light, but even under the best conditions imaginable they would probably stimulate or destroy several hundreds of genes at a time.

Until we can force mutations in some such way as this we can only alter the hereditary composition of ourselves, plants, and animals by combining in one organism genes present in several, and so getting their combined effect. A great deal may thus be done with man. We know very little about human heredity as yet, though about hardly any subject are more confident assertions made by the half-educated; and many of the deeds done in America in the name of eugenics are about as much justified by science as were the proceedings of the Inquisition by the gospels.

The first thing to do in the study of human heredity is to find characters which vary sharply so as to divide mankind definitely into classes. Most ordinary characters are no good for this purpose. We find every gradation of height, weight, hair, and skin colour. A few characters have been found, such as two which determine whether it is safe to transfuse blood from one man into another, which are definitely present or absent, and admit of no doubt. These are inherited in a very simple manner, and divide mankind into four classes.

Now, if we had about fifty such characters, instead of two, we could use them, by a method worked out on flies by Morgan of New York and his associates, as landmarks for the study of such characters as musical ability, obesity, and bad temper. When a baby arrived we should have a physical examination and a blood analysis done on him, and say something like this: 'He has got iso-agglutinin B and tyrosinase inhibitor J from his father, so it's twenty to one that he will get the main gene that determined his father's mathematical powers; but he's got Q4 from his mother, to judge from the bit of hair you gave me (it wasn't really enough), so it looks as if her father's inability to keep away from alcohol would crop up in him again; you must look out for that.'

When that day comes intelligent people will certainly consider their future spouses' hereditary make-up, and the possibility of bringing off a really brilliant combination in one of their future children, just as now we consider his or her health and education, before deciding on marriage. It is as certain that voluntary adoption of this kind of eugenics will come, as it is doubtful that the world will be converted into a human stud-farm.

The third question can be answered in the negative for certain forms at any rate. Out of over 400 mutations observed in one fly, all but two seemed to be disadvantageous; and they showed no definite tendency in any one direction. But, of course, mutation may be biased in other species. The fourth question is largely a matter of mathematics. No competent biologist doubts that evolution and natural selection are taking place, but we do not yet know whether natural selection alone, acting on chance variations, will account for the whole of evolution. If it will, we shall have made a big step towards understanding the world; if it will no more account for all evolution than, for example, gravitation will account for chemical affinity, as was once believed, then biologists have a bigger job before them than many of them think. But a decision of this question one way or the other will greatly affect our whole philosophy and probably our religious outlook.

To turn now to the study of the single animal or plant, physiological researches fall into several classes according to the methods used. Some of us measure the production of small amounts of heat or electrical energy with complicated apparatus, others hunt down unknown chemical substances, or measure accurately the amount of already known ones in the tissues. Taking the biophysicists first, a whole new field has been opened up by recent work on radiation. When X-rays were first applied to living tissues, it was very difficult to get the same result twice running. But now, thanks to the work of our physical colleagues, we can get X-rays of a definite wavelength and intensity, and our results are correspondingly more intelligible. In the same animal one tissue is more sensitive than another to rays of a given wavelength. Moreover, cells are generally more easily upset when engaged in division than at other times. These facts account for our occasional success with X-rays against cancer, and our hope for greater things in the future. It is quite possible that some combinations of invisible wavelengths may be found to have special properties, just as a mixture of red and violet spectral lights gives us the sensation of purple , which intermediate wavelengths do not.

Similarly, sunlight, besides warming us and enabling us

to see, gives us bronzed skin, blisters us, wards off rickets, and cures many cases of tuberculosis. But are all these effects due to the same group of rays acting in the same way? We treat skin tuberculosis with ultra-violet light. Can we increase the curative effect without increasing the danger of severe sunburn? These questions are being answered as I write. The application of rays will gradually be taken out of the doctor's hands. He will write out a prescription, and we will go round to the radiologist's shop next door to the chemist's and ask for the prescribed treatment in his back-parlour. The next man at the counter will be after an apparatus to radiate the buds of his rose-bushes during the winter, and kill off insect eggs which are out of reach of chemicals, without hurting the plants. The quack is already in the market with lamps producing radiation to cure rheumatism and make your hair grow. These are mostly harmless, though a few may be of value; but probably the sale of X-ray tubes, which may cause cancer, will some day be as carefully regulated as that of strychnine.

Physical methods are also being applied in the study of the nervous system. We have by now gone most of the way in the localization of function there, for although a given area of the brain is always concerned in moving the hand, yet a given point in it may cause different movements at different times; just as any one telephonist in an exchange can only ring up certain subscribers, but yet has a fairly wide choice. So we have now got to work out the detail of the processes of excitation and inhibition, as calling up and ringing off are technically called. This involves very accurate measurement of the electrical changes in nerve fibres under different circumstances. Here we are still in the graph and table stage, but probably only about ten years off a fairly comprehensive theory of how the different parts of the nervous system act on one another. This will at once react on psychology, and more slowly on normal life and practical medicine. A great deal that passes as psychology is really rather bad physiology dressed in long words, and the alleged physiology in psychological textbooks is their worst part. We shall alter that. Until, however, we have got a sounder neurology, scientific psychology, except of a fragmentary character, is no more possible than was physiology until chemistry and physics had reached a certain point. And until psychology is a science, scientific method cannot be applied in politics.

In chemical physiology we are after two rather different things. The first is to trace out the chemical processes in the cells, the nature, origin, and destiny of each substance in them. The second, which is much easier, is to trace the effect on cell life of various chemical substances; including those which are normally found in the body, and unusual ones, such as drugs and poisons. The first, if pushed to its logical conclusion, would give us a synthetic cell, and later a synthetic man, or 'robot'. The second would give us a complete system of medicine, which is more immediately required. But, of course, the two react on one another and are not wholly separable.

At the moment the study of cell chemistry is leading to the most interesting results in the case of simple organisms such as yeasts and bacteria. For example, Neuberg of Berlin worked out a number of the steps in the transformation of sugar into alcohol and carbon dioxide by yeast; and was able, by appropriate chemical methods, to side-track the process so that it yielded other products. One of these is glycerine. During the war the Germans were unable to import the fats and oils from which glycerine is generally made. They needed glycerine for their propellent explosives, which contain nitro-glycerine. By getting yeast to make it from sugar they were able, in spite of the blockade, to produce all the nitro-glycerine they wanted.

This special process does not pay in peacetime, but there

are others which do; and every day moulds and bacteria are playing a more important part in industrial chemistry. Similarly, we are now studying the chemical processes in bacteria as carefully as we do those in our own bodies. There is generally a weak link in such a chain; for example, in human beings the links whose breaking gives us diabetes or rickets. If we study the tubercle bacillus carefully we may find his weak point. The relatively direct methods which give us the cure for syphilis are here no use, for the tubercle bacillus is a far tougher organism than the spirochaete, and we cannot yet kill him without killing his host. Similarly, we are trying to find out how the chemical processes in normal and cancerous cells differ.

In man the study of what our body cells can and cannot do is gradually leading us to the perfect diet. It is becoming quite clear that faulty diet gives us some diseases, including most of our bad teeth, and predisposes us to others; and that nothing out of a tin or package so far comes up to natural foodstuffs. On the other hand, as the population of large cities cannot get these, we have got to determine what can be done to improve a diet based largely on milled cereals and tinned milk and meat. It is a tough problem, and for every pound we can spend on research and publicity put together the food-faking firms have a thousand for advertising of 'scientific' foods.

To turn now to the chemical co-ordination of the body, we know that various organs secrete into the blood substances (often called hormones) which profoundly affect the rest of the tissues. A number of these have been obtained in a fairly concentrated form—that is to say, mixed with perhaps only ten or a hundred times their weight of other substances. Only two have been obtained entirely pure, though presumably all will be. Now, if we take one of the most widely popularized of recent therapeutic methods, the grafting of apes' testicles into old or prematurely senile men, this is just an attempt to get a hitherto unisolated hormone into the bloodstream. The operation is expensive, the idea unpleasant, and the graft generally dies in a few years at most. The problem is to isolate the hormone free from other poisonous substances found in most tissue extracts, and later to find its chemical formula and synthesize it. One of the corresponding substances found in the female sex has been obtained free from harmful companions by Allen and Doisy in America.

When we have these substances available in the pure state we ought to be able to deal with many departures from the normal sexual life, ranging from gross perversion to a woman's inability to suckle her children; since lactation, as well as the normal instincts, appears to depend on the presence of definite substances in the blood. We shall also probably be able, if we desire, to stave off the sudden ending of woman's sexual life between the ages of forty and fifty. It is worth pointing out that there is no serious reason to believe that any of the rather expensive products of the sex glands now on the market, and often prescribed by doctors, are of any value except as faith cures.

A much more ambitious attempt to deal with old age is being started by Carrel. Cultures from individual cells from a chicken can be kept alive in suitable media for twenty years, and as far as we know for ever. To live they must have certain extracts of chicken embryo. The blood of a young fowl contains substances (which can be separated by suitable methods) that both stimulate and check their growth. The former is absent in very old fowls. The problem of perpetual youth has, therefore, been solved for one kind of cell. But to make a pullet immortal we should have to solve it for all the different cells of its body at once. We do not know if this is possible, or whether it is like trying to design a society which is ideal alike for cowboys, automobile manufacturers, and symbolist poets, all of whom can hardly flourish side by side. Fifty years hence we shall probably know whether it is worth seriously trying to obtain perpetual youth for man by this method. A hundred years hence our great-grandchildren may be seeing the first results of such attempts.

Besides these rather sensational substances which were first detected by their effects on organs, the proper working of the organism depends on the amount of quite well-known bodies, such as sugar, oxygen, and lime in the blood. We are gradually getting to know the amounts of these required for health, but it is much harder to estimate the amount needed of such a substance as, say, insulin. We can now kill an animal and produce a fluid from inorganic constituents that will keep its heart or liver alive for a day or more. Soon it will be a matter of months or years. To keep tissues alive for a time comparable with the life of their owner we shall have to have about 100 substances, but perhaps not very many more, present in the normal amounts in the fluid perfusing them. At present we only know the correct quantity of some twenty, if that. Given this knowledge and the means of applying it, we could make good the deficiency of any organ but the nervous system. We could grow human embryos in such a solution, for their connection with their mother seems to be purely chemical. We could cut our beefsteak from a tissue culture of muscle with no nervous system to make it waste food in doing work, and a supply of hormones to make it grow as fast as that of an embryo calf.

In pharmacology our knowledge rests mainly on a series of lucky accidents. A few of the complicated substances made by plants have a striking effect on animals, but why a molecule of a given build has a given physiological effect we are only beginning to discover. When we know, we should be able to make as great an advance on plant products as we did with dyes when the relations between colour and chemical composition were discovered. If we had a drug that was as good a pain-killer as morphine, but one-tenth as poisonous and not a habit former, we could use it indiscriminately; and wipe out a good half of the physical pain in human life at one stroke.

Such are a few of the possibilities of our science. It is easy enough to say what we would do if we had a method to measure *A* or isolate *B*. But it is in inventing and applying these methods that our biggest problems often arise.

WHEN I AM DEAD

WHEN I am dead I propose to be dissected; in fact, a distinguished anatomist has already been promised my head should he survive me. I hope that I have been of some use to my fellow creatures while alive, and see no reason why I should not continue to be so when dead. I admit, however, that if funerals gave as much pleasure to the living in England as they do in Scotland I might change my mind on the subject.

But shall I be there to attend my dissection or to haunt my next of kin if he or she forbids it? Indeed, will anything of me but my body, other men's memory of me, and the results of my life, survive my death? Certainly I cannot deny the possibility, but at no period of my life—least of all during the war, when I was nearest to death—has my personal survival seemed to me at all a probable contingency.

If I die as most people die, I shall gradually lose my intellectual faculties, my senses will fail, and I shall become unconscious. And then I am asked to believe that I shall suddenly wake up to a vivid consciousness in hell, heaven, purgatory, or some other state of existence.

Now, I have lost consciousness from blows on the head, from fever, anaesthetics, want of oxygen, and other causes; and I therefore know that my consciousness depends on the physical and chemical condition of my brain, and that very small changes in that organ will modify or destroy it.

But I am asked to believe that my mind will continue without a brain, or will be miraculously provided with a new one. I am asked to believe such an improbable theory on three grounds.

The first set of arguments are religious. My Catholic friends hope to survive death on the authority of the Church; some of my Protestant acquaintances rely on the testimony of the Bible. But they do not convince me, for the Church has taught doctrines which I know to be false, and the Bible contains statements—for example, concerning the earth's past—which I also know to be false.

Other Christians ask me to believe in immortality on the authority of Jesus. That is a much more cogent argument, because Jesus was not only a great man and a great ethical teacher, but a great psychologist. But the characteristic part of any man's teaching is what is novel and heretical in it, and not what he and his audience take for granted.

Now Jesus, and the Pharisees, at any rate, among his hearers, took the resurrection of the dead for granted. They also took for granted that madness was due to possession by devils.

When Jesus tells me to love my enemies he is speaking his own mind, and I am prepared to make the attempt; when he tells me that I shall rise from the dead he is only speaking for his age, and his words no more convince me of immortality than of demoniacal possession.

Again, I am told that men have always believed in immortality, and that religion and morality are impossible without it. The truth of this ludicrous statement may be tested by referring to the first seven books of the Bible. They are full of religion and ethics, but contain no reference to human survival of death. Nor did the Psalmist believe in it. 'The dead praise not Thee, O Lord,' he said, 'neither all they that go down into silence.'

As a matter of fact, the belief in a life beyond the grave reached its culminating point in Egypt four or five thousand years ago, when the rich, at any rate, seem to have spent more money in provision for their future life than for their present. To judge from what has come down to us of his writings, Moses, the Man of God, who was well versed in Egyptian religion, had no more use for a future life than for the worship of crocodiles.

The belief in personal immortality seems to have spread gradually out from Egypt, along with the use of copper, bronze, and gold, and often, especially in Polynesia, accompanied by the practice of mummification. It is an attractive doctrine, and is only now beginning to lose its grip on the human mind. But it is losing it.

Most of the men under my command whom I got to know during the war believed in God. But I think the majority thought that death would probably be the end of them, and I am absolutely certain that that is the view of most highly educated people.

I am also asked to believe in a future life on philosophical grounds. There are a number of arguments which seem to prove that my soul is eternal and indestructible. Unfortunately, they also prove that it has existed from all eternity. And, though I am quite willing to believe that ten years after my death I shall be as I was ten years before my birth, the prospect cannot be said to thrill me.

Again, it is argued that justice demands a future life in which sin will be punished, virtue rewarded, and undeserved suffering compensated. Such a view would, of course, involve a future life for animals, in which the hunted hare would demand of its pursuers for what crime it was torn to pieces.

But it assumes that the universe is governed according to human ideas of justice. The sample of it which we know is certainly not so governed; and I see no reason to suppose that its inequalities are redressed elsewhere. No doubt I should like to see them redressed, but then I should like to see England a land fit for heroes to live in, though I do not suppose that I will.

As a matter of fact, conditions in the present world have

been improved largely by recognizing that the laws governing it are not the laws of justice but the laws of physics. As long as people thought that cholera epidemics were a punishment for the people's sins they continued. When it was found that they were due to a microbe they were stopped.

We do ourselves no good in the long run by telling ourselves pleasant fairy-tales about this world or the next. If we devoted the energy that we waste in preparing for a future life to preventing war, poverty, and disease, we could at least make our present lives very satisfactory for most people, and if we were happy in this world we should not feel the need of happiness hereafter.

It is worth remembering how very few people in the past have believed in the justice of the universe. Most Christians have believed that unbelievers and unbaptized children were doomed to spend eternity in hell, while Buddhists believe that all existence is an evil. It is only our own perhaps unduly optimistic age that has assumed that a future life is to be desired.

The arguments of the spiritualists, theosophists, and their like claim to be based on evidence of a kind which appeals to the scientific mind, and a dozen or so distinguished men of science have been spiritualists. Some of this evidence is based on fraud. The bulk of it proves nothing. I have often taken part in the receipt of messages alleged to come from spirits, but they have never given any verifiable information unknown to any members of the circle.

Even, however, if we accepted the view of the spiritualists that a medium can somehow get into communication with the mind of a dead man, what would this prove? If we accept spiritualism we must certainly accept telepathy. Now, I can see little more difficulty in two minds communicating across time than across space.

If I can transmit thoughts to a friend in Australia today,

that does not prove that my mind is in Australia. If I give information to a medium in the year 1990, ten years after my death, that will not prove that my mind will still be in existence in 1990.

To prove the survival of the mind or soul as something living and active we should need evidence that it is still developing, thinking, and willing; spiritualism does not give us this evidence. Shelley is said to have dictated a poem to a medium. It was a very bad poem. Nor do the post-mortem productions of Oscar Wilde reach the standard which he attained when alive.

The accounts given by spirits of a future life vary from land to land and from age to age. Medieval ghosts generally come from purgatory, like Hamlet's father; more rarely from heaven or hell. Hindu and Buddhist ghosts are awaiting their next incarnation. Modern European spirits generally profess a rather diluted Christianity. In fact, the evidence as to the nature of a future life is so contradictory that we must in any case reject most of it.

Personally, I think that all accounts of a future life are mere reflections of the medium's own opinions on the subject, which are of no more value than anyone else's. (The possible interpretation of spiritualistic phenomena which I have given is one which has commended itself to some of their most careful investigators, but has obtained little publicity as it has no emotional appeal.)

But if death will probably be the end of me as a finite individual mind, that does not mean that it will be the end of me altogether. It seems to me immensely unlikely that mind is a mere by-product of matter.

For if my mental processes are determined wholly by the motions of atoms in my brain I have no reason to suppose that my beliefs are true. They may be sound chemically, but that does not make them sound logically. And hence I have no reason for supposing my brain to be composed of atoms.

In order to escape from this necessity of sawing away the branch on which I am sitting, so to speak, I am compelled to believe that mind is not wholly conditioned by matter. But as regards my own very finite and imperfect mind, I can see, by studying the effects on it of drugs, alcohol, disease, and so on, that its limitations are largely at least due to my body.

Without that body it may perish altogether, but it seems to me quite as probable that it will lose its limitations and be merged into an infinite mind or something analogous to a mind which I have reason to suspect probably exists behind nature. How this might be accomplished I have no idea.

But I notice that when I think logically and scientifically or act morally my thoughts and actions cease to be characteristic of myself, and are those of any intelligent or moral being in the same position; in fact, I am already identifying my mind with an absolute or unconditioned mind.

Only in so far as I do this can I see any probability of my survival, and the more I do so the less I am interested in my private affairs and the less desire do I feel for personal immortality. The belief in my own eternity seems to me indeed to be a piece of unwarranted self-glorification, and the desire for it a gross concession to selfishness.

In so far as I set my heart on things that will not perish with me, I automatically remove the sting from my death. I am far more interested in the problems of biochemistry than in the question of what, if anything, will happen to me when I am dead.

Until this attitude is more general the latter question will remain too charged with emotion to make a scientific investigation of it possible. And until such an investigation is possible a man who is honest with himself can only answer, 'I do not know.'

SCIENCE AND THEOLOGY AS ART-FORMS

RELIGION and science are human activities with both practical and theoretical sides. There is at present a certain degree of conflict between them, and this will undoubtedly continue for some generations. During this conflict the disputants have tended to emphasize the differences between them. But their resemblances are equally interesting, and perhaps throw a good deal of light on the differences.

It is only very recently that they have had a chance of diverging. Readers of the Pentateuch, or of the contemporaneous or earlier religious literature of Egypt or Mesopotamia, will find it very difficult to disentangle the science from the religion. The Pentateuch contains some very good applied science in the sanitary laws of Moses. The palaeontology of Genesis is also correct in many points; particularly in describing a period of the earth's history before the origin of life, followed by the appearance of animals in the seas, and only later on land, man being the last creation. It is of course wrong in putting the origin of plants before that of stars, of birds before that of creeping things, and in several other respects.

Unfortunately, however, since Moses' time science and religion have diverged, not without a certain loss to both. The reason for this divergence can best be seen when we study some typical scientific and religious minds at work. Each starts from a certain experience, and builds up a system of thought to bring it into line with the remainder of experience. The organic chemist says, 'The substance I have just made is a liquid with a characteristic smell, melting at

31 °C, boiling at 162 °C, and whose compound with phenyl-hydrazine melts at 97 °C. I have probably synthesized furfural in a new way.' The saint says, 'I have had an experience very wonderful and rather difficult to describe in detail, but I interpret it to mean that God desires me to devote myself to preaching rather than to shut myself up from the world.'

The scientific man, then, starts from experiences in themselves emotionally flat, though to him perhaps interesting enough. He may end by producing a theory as exciting as Darwinism, or a practical invention as important as antiseptics or high explosives. The mind of the religious man on the other hand works on a descending scale of emotions. The dogma, prophecy, or good works which he may produce are inevitably less thrilling than his religious experience.

It is more interesting to most minds to read or speculate about the distances of the stars than to measure the positions of their images on a photographic plate. But it is less interesting to read a work on justification by faith or on St Thomas's theory of transubstantiation than to take part in a well-conducted church service.

Now, the rather dull raw material of scientific thought consists of facts which can be verified with sufficient patience and skill. The theories to which they have given rise are far less certain. They change from generation to generation, even from year to year. And the religious opponents of science tend to scoff at this perpetual change. In scientific thought we adopt the simplest theory which will explain all the facts under consideration and enable us to predict new facts of the same kind. The catch in this criterion lies in the word 'simplest'. It is really an aesthetic canon such as we find implicit in our criticisms of poetry or painting. The layman finds such a law as $\partial x/\partial t = \varkappa(\partial^2 x/\partial y^2)$ less simple than 'it oozes', of which it is the mathematical statement. The physicist reverses this judgement, and his statement is certainly the more fruitful of the two, so far as prediction is concerned. It is however, a statement about something very unfamiliar to the plain man, namely, the rate of change of a rate of change. Now, scientific aesthetic prefers simple but precise statements about unfamiliar things to vaguer statements about well-known things. And this preference is justified by practical success. It is more satisfactory scientifically to say: 'The blood-vessels in John Smith's skin are dilated by a soluble toxin produced by a haemolytic streptococcus growing in his pharynx', than to say that he has scarlet fever. It suggests methods of curing and preventing the disease. But only a few people have seen a streptococcus, and no one has seen a toxin in the pure state. In physics, the most developed of the sciences, things have gone so far that many physicists frankly say that they are describing atom models and not atoms. Atoms themselves have the same sort of reality as chairs and tables, because single atoms can be seen if going fast enough. But when we come to their internal structure we can only say that they behave, in some important respects, as if electrons were going round in them with such and such velocities in such and such orbits. If that is the real structure we can calculate the velocities with a great deal more accuracy than that with which a speedometer gives the speed of an automobile; and verifiable predictions based on these calculated speeds come out with very great accuracy. But the speeds are not observable, and physicists are becoming less and less careful as to what hypothetical events they postulate to explain observable phenomena, provided the hypotheses enable them to predict accurately.

Einstein showed that we could explain and predict slightly better if he substituted other conceptions for those of space and time, and his own substitutes will doubtless be replaced in their turn. Now, if Einstein is right, or even partly right, no physicists before his time knew quite what they were talking about when they used the ideas of distance and time, and practically every statement that they made which purported to be accurate was false. So presumably is every such statement of a modern physicist. Similarly, chemists supposed that the weight of a chlorine atom was 35.46 units until Aston showed that chlorine atoms were a mixture of two kinds whose weights are 35 and 37. Almost all the deductions from the premise were right, but the premise itself was wrong. I have no doubt that biological theory is equally riddled with falsehoods.

In fact, the experience of the past makes it clear that many of our most cherished scientific theories contain so much falsehood as to deserve the title of myths. Their claims to belief are that they contradict fewer known facts than their predecessors, and that they are of practical use. But there is one very significant feature of the most fully developed scientific theories. They tell us nothing whatever about the inner nature of the units with which they deal. Electrons may be spiritually inert, they may be something like sensations, they may be good spirits or evil spirits. The physicist, however, can only tell us that they repel one another according to a certain law, are attracted by positive charges according to another law, and so on. He can say nothing about their real being, and knows that he cannot.

It is, I suppose, the fact that there is no great stability to be found in scientific theories which leads the opponents of science to talk of the intellectual bankruptcy of our age; generally as a preliminary to adopting beliefs current in medieval Europe, India, or Bedlam. We must therefore proceed to examine the claims of the theological beliefs which are offered as substitutes or supplements for science. Religious and moral experience are facts. Most people can obtain a certain amount of religious experience by a very moderate effort. It would be ridiculous not to interpret it.

But its interpretation is still in a pre-scientific stage. There is something true in theology, because it leads to right action in some cases, and serves to explain certain otherwise difficultly explicable facts about the human soul. There is something untrue, because it often leads to wrong actions and leaves a great deal which is within its province, for example the origin of evil, unexplained.

In India and in pre-Christian Europe theology developed gradually. Gods or devils could be postulated as required. This, however, led to a polytheism incompatible either with the unity of nature or the unity of duty. Since about AD 400, however, Christian theology has altered very little. But the moral consciousness of Christians has altered a great deal. For example, it now condemns slavery and cruelty to animals. St Paul condoned the one and ignored the other. But it is alleged by theologians that our moral consciousness is one source of our knowledge of God. If the one develops, so should the other. Now, in the early or growing stages of a religion a large number of myths are produced which interpret the moral and religious consciousness of the time. Jesus habitually used vivid imagery. Sometimes he was obviously speaking in parables. Sometimes a doubt exists, as with regard to the words used at the Last Supper. If taken as a statement of fact, they lead to a belief in transubstantiation. If not, they are merely an expression of solidarity like 'I am the vine, ye are the branches.' Some of his other statements are taken literally by all Christian Churches, and it is claimed that a belief in them is a more or less essential prerequisite to a Christian life. This appears to be doubtful.

But where Jesus used parables as far as possible, his followers stated innumerable doubtful propositions as facts, and attached more importance to a belief in them than the followers of the non-Christian religions have attached to similar statements. Some of these propositions were about the structure of the universe, for example 'The Holy Ghost proceeds from the Father and the Son', others about events, for example 'Jesus descended into Hell.' Scientific men clearly cannot complain of the theologians for making myths. They do it themselves. Sometimes they take them very seriously. Occasionally the public does so. The ether is a highly mythical substance filling empty space, for which relativistic mechanics have little use. But broadcasting has made it a popular myth. Today probably more men in England believe in it than in Jesus's bodily ascension to heaven.

The main objection to religious myths is that, once made, they are so difficult to destroy. Chemistry is not haunted by the phlogiston theory as Christianity is haunted by the theory of a God with a craving for bloody sacrifices. But it is also a fact that while serious attempts are constantly being made to verify scientific myths, religious myths, at least under Christianity and Islam, have become matters of faith which it is more or less impious to doubt, and which we must not attempt to verify by empirical means. Chemists believe that when a chemical reaction occurs, the weight of the reactants is unchanged. If this is not very nearly true, most of chemical theory is nonsense. But experiments are constantly being made to disprove it. It obviously cannot be proved, for, however accurately we weigh, the error may still be too small for us to observe. Chemists welcome such experiments and do not regard them as impious or even futile.

Christians almost all believe that prayer is sometimes answered. Now, all prayers are not answered, and desired events often occur for which no one has prayed. Hence individual instances of answered prayers are useless. We must have statistical evidence. It has been proposed from time to time that a group of believers should pray for the recovery of the patients in one wing of a hospital over a period of some months, and the number of deaths in it be compared with that in the other wing. The experiment has always been refused, partly on the ground that 'Thou shalt not tempt the Lord thy God', partly through lack of faith. Until it has been made, I do not propose to ask for the prayers of any congregation on my behalf.

In the absence of experimental evidence Galton attempted a statistical investigation. He considered that of all classes of society in England those most prayed for were the sovereigns and the children of the clergy. If prayer is effective they should live appreciably longer than other persons exposed to similar risks of death. So kings were compared with lords, and the children of the clergy with those of other professional men. The conclusion to which his numbers led was that these much-prayed-for persons had slightly shorter lives than those with whom he compared them. The difference was not, however, great enough to make it probable that prayers have any harmful effect.

On the other hand, there can be little doubt that either one's own prayers or the knowledge that others are praying for one may serve as a safeguard against temptation, or effect the complete or partial cure of functional diseases of the nervous system. There is therefore, at any rate, a certain truth in the efficacy of prayer; but this efficacy may be explained on psychological grounds, and does not necessarily imply a divine interference in the order of nature.

Nevertheless, religious experience is a reality. It cannot be communicated directly, but those who experience it can induce it in others by myth and ritual. This fact is most fully recognized by Hindus. The simpler of them believe in a vast and complex system of myths, and attempt to gain their personal ends by placating various deities. The more intellectual do not pretend that myths are true, or that the ritual has any effect other than a psychological influence on the participants. A well-known Hindu mathematician, who would sooner have died than eaten beef, once asked a British colleague whether it would cause real surprise in England if the Archbishop of Canterbury were to deny the existence of God in the House of Lords. In India, it appears, the corresponding event would excite no more comment than does a denial of the existence of the soul by a professor of psychology, or of the reality of matter by a professor of physics, in England or America.

This essay is emphatically not a defence of Hinduism, which has excused every kind of evil, from murder and the prostitution of young girls to self-mutilation and refusal to wash. It is, moreover, the chief prop of a system of hereditary class distinctions based in part on differences of colour; by contrast with which Louis XIV appears an equalitarian and Pizarro a champion of racial equality. And I find many of its myths disgusting. But the fact remains that it has lasted a great deal longer than Christianity and shows far fewer signs of decay. It has more than half as many adherents, and on the whole affects their lives to a greater extent.

One of the gravest errors into which the Christian religion has fallen is the view that a myth or dogma cannot influence you unless you believe it. Christians in general believe in the existence of St Bartholomew and St James the Less, but they are much less influential people at the present day than Cordelia and Father Christmas, in whom very few adults believe. If the Christian Churches continue to make belief the test of religion, it appears to me that one of three things will happen.

If they maintain their influence they will sterilize scientific thought, and either slow down human progress or render their adherents as defenceless against non-Christian races armed with science as were the Asiatics against the Christian races during the nineteenth century. The Caliph

Mutawakkil, who, perhaps more than any other man, assured the triumph of orthodoxy and the suppression of independent thought in the Muhammadan world, was responsible for its conquest by the Christians, when these latter were allowed to think for themselves about the nature of matter and hence to produce steam-engines and high explosives. The Christian Churches are preventing people from thinking for themselves about life. In the interests of theological orthodoxy Christian children may not be taught about evolution. In those of morality and decency they may not be taught how their bodies work. If any Asiatic people begins as a whole to think biologically before those of European origin do so, it will dominate the world, if the lessons of the past are any guide to the future.

It is perhaps more probable that under such circumstances religion will decline and become unimportant in human life. This would, perhaps, be a misfortune, for it is probably the essence of religion that a man should realize that his own happiness and that of his neighbours are not his only concerns. If we abandon religion completely we shall probably deify ourselves and behave in a wholly selfish manner, or deify the State and behave as the more violently patriotic Germans did in the late war. (And as the war continued the other nations conformed more and more to the German model.)

It may be that religion as we know it will some day be superseded altogether, for many people nowadays lead good lives without it. It is possible to become convinced on philosophical grounds of the supreme reality and importance of the spiritual, without postulating another world or even a personal God. And such convictions may be supported by mystical experience. However, it is improbable that most people are capable of the abstraction necessary if such a point of view is to be adopted. It is even harder, though of course possible, to worship one's own moral convictions, as Mr Bertrand Russell does, while believing that they are an unimportant by-product of a universe which, as a whole, is indifferent to them. One of these views of the world may well be true, but I do not believe that most people can attain to it for some centuries. And self-worship and State-worship are within the reach of all.

A third possibility is the rise on the ruins of Christianity of a religion with a creed in harmony with modern thought, or more probably the thought of a generation ago. Traces of such a creed may be found in the utterances of prominent spiritualists, in the economic dogmas of the communist party, in the writings of believers in creative evolution, and elsewhere. A new religion would crystallize the scientific theories of its own age. The old religions are full of outworn science, including the astronomical theory of a solid heaven, the chemical theory that water, bread, books, and other objects can be rendered holy by special processes, and the physiological theory that a substance called a soul leaves the body at the moment of death. I remember hearing the headmaster of Eton informing the school from the pulpit that the body lost weight (or gained it, I forget which) at the moment of death. This materialistic view of the nature of the soul is quite prevalent in Christian circles. A new religion would probably include in its creed the reality of the ether, the habitability of Mars, the duty of man to co-operate in the process of evolution, the existence of innate psychological difference between the human races, and the wickedness of either capitalism or socialism.

Now, Christianity has ceased to obstruct astronomy and geology with such texts as 'He hath established the world so fast that it cannot be moved', and 'The waters under the earth'. It will find it a harder task to admit that it has been as wrong about the nature of the soul as about the nature of heaven. Nevertheless, such an admission will have to be made if Christianity is to survive as the religion of any appreciable fraction of educated men and women. The experience of the past shows that such an admission is not impossible. In the words of Renan, 'Si le parti radical, parmi nous, était moins étranger à l'histoire religieuse, il saurait que les religions sont des femmes dont il est très facile de tout obtenir, si on sait les prendre, impossible de rien obtenir, si on veut procéder à haute lutte.'

On the other hand, the dogmas of a new religion would include enough of contemporary science to be fairly plausible, and enough of contemporary ethics to be fairly practicable. It would thus constitute a far more serious obstacle than Christianity to scientific and ethical progress in the future, even if it led to a momentary advance in scientific and ethical education.

I therefore suggest the following standpoint for consideration. Religious experience is a reality. Hence it affords a certain insight into the nature of the universe, and must be considered in any account of it. But any account given of religious experience must be regarded with the gravest suspicion. Some of the greatest mystics described their experience in language which is frankly and admittedly self-contradictory. Such expressions as 'a delicious desert', 'a dazzling darkness', 'ein lauter Nichts', 'it neither moves nor rests', are common in their writings. Others have admitted their ignorance of details. 'Nescio, nescio, quae jubilatio, lux tibi qualis' ('I know not, I know not, what is thy joy and thy light'), said Bernard of Cluny. Where a more definite statement is made, it is even less reliable than the crude interpretation of our sense-data. Our ordinary perceptions tell us that the sun, a body of indeterminate size, but certainly not as large as a country, let alone the whole world, rises every morning from a level below our own and descends again at evening. I suggest that the data of religious experience as to the nature of God and his relation to man are no more reliable than those of crude perception concerning the sun and its relation to the earth. In fact, they are less so, for different mystics disagree with one another. Nevertheless, mystical experience may be as capable of scientific investigation and explanation as sensuous experience, though the process would be more difficult because the experience is rarer. Such an interpretation, provided its results did not contradict those of other branches of knowledge, would constitute for the first time a theology worthy of the name of a science. Whether its conclusion would be theistic or not it is impossible to judge.

The theology, or rather theologies, of today are something quite different. Religious experience has so far been described mainly in symbols drawn from everyday life. Some such symbolic expression seems to be inevitable. Unfortunately, the vast majority of mystics, or at least their disciples, have come to take the symbols seriously. Some modern religious literature, however, furnishes a most gratifying exception to this rule. The writings of L. P. Jacks, for example, which undoubtedly sprung from a considerable body of religious experience, are mainly cast in the form of fiction. His account of a future life in *All Men are Ghosts* is the most attractive that I have so far come across, but it does not purport to be accurate. The similar accounts given by the saints are also mostly, if not all, based on genuine religious experience, but they are not for that reason to be believed. Theology or mythology is an art-form used to express religious experience. Both theology and scientific theory may be valuable guides to conduct, as well as beautiful in themselves, but that does not make them true. The theory that heat is a substance, not a mode of motion, enabled Watt to make the calculations on which the design of his steam-engines was based. The calculations were right, but the theory was wrong. The Christian saints believed God to be a person (though the Buddhists did not). Their lives were often, though not always, admirable; and their religious experience on the whole conformed with their beliefs. But it does not follow that their beliefs were correct.

If such a point of view is adopted, the literature of Christianity will come to be regarded as of mainly symbolical value; but yet as showing forth a real experience which could perhaps have been expressed in no other way, at the time when it was composed. An even larger proportion of the sayings of Jesus will be regarded as parables and therefore to be interpreted; rather than dogmas to be believed, and then fitted, not without strain, to the rest of experience. Christians will learn to take many of the Churches' prohibitions no more seriously than St Paul's veto on things strangled. They will regard their institutions rather as tokens of solidarity with the past and the future than as means of salvation. And they will rank theology with poetry, music, rituals, architecture, sculpture, and painting, as an expression of religion, but not its essence.

Perhaps a summary of the ideal relationship of religion and science would be somewhat as follows. Religion is a way of life and an attitude to the universe. It brings men into closer touch with the inner nature of reality. Statements of fact made in its name are untrue in detail, but often contain some truth at their core. Science is also a way of life and an attitude to the universe. It is concerned with everything but the nature of reality. Statements of fact made in its name are generally right in detail, but can only reveal the form, and not the real nature, of existence. The wise man regulates his conduct by the theories both of religion and science. But he regards these theories not as statements of ultimate fact but as art-forms.

THE LAST JUDGEMENT

> Denique montibus altior omnibus ultimus ignis
> Surget, inertibus ima tenentibus, astra benignis,
> Flammaque libera surget ad aëra, surget ad astra,
> Diruet atria, moenia, regna, suburbia, castra.
>
> BERNARD OF CLUNY,
> *De contemptu mundi*, Book I

THE star on which we live had a beginning and will doubtless have an end. A great many people have predicted that end, with varying degrees of picturesqueness. The Christian account contains much that is admirable, but suffers from two cardinal defects. In the first place, it is written from the point of view of the angels and a small minority of the human race. The impartial historian of the future could legitimately demand a view of the communiqués of the Beast of the Book of Revelation and his adherents. For, after all, the Beast and his false prophet could work miracles of a kind, and were admittedly able propagandists. So perhaps 'Another air raid on Babylon beaten off. Seventeen archangels brought down in flames' might record some of the earlier stages in the war, while 'More enemy atrocities. Prophet cast into burning sulphur' would chronicle the peace terms.

But the more serious objection is perhaps to the scale of magnitudes employed. The misbehaviours of the human race might induce their creator to wipe out their planet, but hardly the entire stellar system. We may be bad, but I cannot believe that we are as bad as all that. At worst our earth is only a very small septic area in the universe, which could be sterilized without very great trouble, and conceivably is not even worth sterilizing.

I prefer *Ragnarok*, the Doom of the Reigners, which closes the present chapter in world history according to Norse mythology. Here mankind perish as an episode in a vaster conflict. It is true that they misbehaved first.

> Hart es i heimi, hordomr mikkil,
> Skeggi-aold, skalm-aold, skildir klofnir,
> Vind-aold, varg-aold, aðr vaerold steipisk[1]
>
> (Hard upon earth then, many a whoredom,
> Sword-age, axe-age, shields are cloven,
> Wind-age, wolf-age, ere world perish),

says the Norse Sybil in the *Volospa*. But human events are a symptom rather than a cause. The gods are to be destroyed by the powers of darkness. Fenri, the wolf, will eat Odin, and actually get the world between his teeth, though he will fail to swallow it. There is a happy ending, probably due to Christian influence. Balder returns from the dead, and rules over the descendants of two survivors of the human race. But one episode is of considerable interest. In the middle of the fight the Sun becomes a mother, and both she and her daughter survive it. In Scandinavia, of course, the Sun, who is kindly but rather ineffective, is a female, a conception impossible to the inhabitants of hotter climates.

Now, fission is one of the vices to which suns are subject. Indeed, something like half the 'fixed' stars known to us are double or multiple. Apparently the reason for splitting is as follows. A star always has a certain amount of angular momentum, or spin, due to its rotation on its axis. As it loses heat it gets smaller, but keeps the same amount of spin. So it has to go round faster, and finally splits in two, like a bursting flywheel, owing to its excessive speed. The Sun certainly does not seem likely to do this, for it turns round

[1] The letter ð was pronounced like th in then, þ like th in thin.

its axis only once in about four weeks; whereas in order to split, it would have to do so once in less than an hour. But we can see only its outside, and last year Dr Jeans, the president of the Royal Astronomical Society, suggested that the Sun's inside might be rotating much faster, and that no one could say that it would not divide tomorrow. Naturally, such an event is rather unlikely. The Sun has gone on for several thousand million years without doing so. But it is apparently possible.

The results for the Earth would be disastrous. Even if the Sun's heat did not increase so greatly as to roast mankind forthwith, the Earth would cease to revolve in a definite orbit, and year by year would approach the pair of suns nearer at one season, retreat from them further at another, while they themselves would gradually separate, and therefore approach nearer to the Earth. Long before a collision occurred we should have come so close to one of them that, under the radiation from a sun covering perhaps a tenth of the sky, the sea would have boiled over and mankind perished.

The Sun might perhaps do several other things. It might cool down, and a generation ago it seemed very plausible that it would do so within a few million years. But as we now know that for the thousand million years or so since the first ice-age recorded by geology it has not got much cooler, there is no reason to suppose that it will begin to do so for a very long time indeed. Modern physics suggests, indeed, that it will shine for at least a million million years. But before that time comes, something very strange, as we shall presently see, will have happened to our own planet.

Stars occasionally burst, expanding enormously, giving out a vast amount of heat, and then dying down again. No one knows why this occurs, but it does seem to happen to stars not at all unlike the Sun. If it happened to the Sun, the Earth would stand as much chance of survival as a butterfly in a furnace. But these explosions are also rare. No star at all near to us has exploded during human history. If Sirius, let us say, exploded in this manner, he would send nearly as much light to the Earth as does the Moon, and would be visible by day. We cannot say whether this kind of ending for our world is likely or not until we know more as to why it happens to other suns than our own.

Others have suggested a comet or some stray heavenly body as a destroyer. Against this we have the fact that on all the continents nothing more than a few miles in diameter can have fallen in the last few hundred million years. The great meteor embedded in the desert in Arizona may have formed part of a comet, and some of the scars on the Moon may be due to collisions with wandering matter. But the improbability of a collision which would desolate any large part of the Earth's surface is enormous, even though the Arizona meteorite would have made a considerable mess of London or New York. It has been suggested that a heavy body passing near the Earth might drag it out of its orbit. The orderly and nearly circular character of the orbits of all the planets round the Sun shows that they have not been greatly perturbed for a very long time, and probably since their formation. One cannot say that they will never be so perturbed, but one can assert that the odds against any such event in the next million years are more than a thousand to one.

All the possibilities that I have catalogued are essentially accidents. Some of them may happen, just as I may be killed in a railway accident; but just as my body will not go on working for ever, apart from any accidents, so the Earth carries with it through space what will certainly alter its conditions profoundly, and very possibly destroy it as an abode of life. I refer to the Moon.

Our Scandinavian ancestors did not neglect our satellite in their account of the twilight of the gods.

Austr byr in aldnar i Iarnviði
Ok foeðir þar Fenris kindir,
Verðr af þeim aollom einar nokkar
Tungls tiugari, i trollz hami.

(Eastward in Ironwood sits the old witch
And breeds Fenri's children,
Of them all one shall be born
Shaped like an ogre, who shall pitch the moon down.)

Now, here the sibyl who described the future to Odin was substantially in agreement with modern astronomy. The Moon will one day approach the Earth so close as to be broken up, and very possibly to destroy the Earth's surface features. Certain Muhammadan theologians have interpreted the first verse of the Sura called 'The Moon', 'The hour is come and the moon is split', as referring to the end of the world. But outside Scandinavia the prophets of doom have generally described the stars as falling out of heaven, which is an impossibility, for the same reason that a million elephants cannot fall on one fly. They are too large.

In what follows I shall attempt to describe the most probable end of our planet as it might appear to spectators on another. I have been compelled to place the catastrophe within a period of the future accessible to my imagination. For I can imagine what the human race will be like in forty million years, since forty million years ago our ancestors were certainly mammals, and probably quite definitely recognizable as monkeys. But I cannot throw my imagination forward for ten times that period. Four hundred million years ago our ancestors were fish of a very primitive type. I cannot imagine a corresponding change in our descendants.

So I have suggested the only means which, so far as I can see, would be able to speed up the catastrophe. The account given here will be broadcast to infants on the planet Venus some forty million years hence. It has been rendered very freely into English, as many of the elementary ideas of our descendants will be beyond our grasp:

'It is now certain that human life on the Earth's surface is extinct, and quite probable that no living thing whatever remains there. The following is a brief record of the events which led up to the destruction of the ancient home of our species.

Eighteen hundred and seventy-four million years ago the Sun passed very close to the giant star 318.47.19543. The tidal wave raised by it in our sun broke into an incandescent spray. The drops of this spray formed the planets, of all of which the Earth rotated by far the most rapidly. The Earth's year was then only very slightly shorter than now; but there were 1,800 days in it, each lasting only a fifth of the time taken by a day when men appeared on Earth. The liquid Earth spun round for a few years as a spheroid greatly expanded at the equator and flattened at the poles by its excessive rotation. Then the tidal waves raised in it by the Sun became larger and larger. Finally the crest of one of these waves flew off as the Moon. At first the Moon was very close to the Earth, and the month was only a little longer than the day.

As the Moon raised large tides in the still liquid Earth the latter was slowed down by their braking action, for all the work of raising the tides is done at the expense of the Earth's rotation. But by acting as a brake on the Earth, the Moon was pushed forward along its course, as any brake is pushed by the wheel that it slows down. As it acquired more speed it rose gradually further and further away from the Earth, which has now a solid crust, and the month, like the day, became longer. When life began on the Earth the Moon was already distant, and during the sixteen hundred million years before man appeared it had only moved away to a moderate degree further.

When these distances were first measured by men the Moon revolved in twenty-nine days, and the braking action of the tides amounted to twenty thousand million horsepower on the average. It is said that the effect of tidal friction in slowing down the Earth's rotation, and therefore lengthening the day, was first discovered by George Darwin, a son of Charles Darwin, who gave the earliest satisfactory account of evolution. However, there is reason to believe that both these personages are among the mythical culture-heroes of early human history, like Moses, Laotzu, Jesus, and Newton.

At this time the effect of tidal friction was to make each century, measured in days, just under a second shorter than the last. The friction occurred mainly in the Bering Sea between northern Asia and America. As soon as the use of heat engines was discovered, man began to oxidize the fossil vegetables to be found under the Earth's surface. After a few centuries they gave out, and other sources of energy were employed. The power available from fresh water was small, from winds intermittent, and that from the Sun's heat only available with ease in the tropics. The tides were therefore employed, and gradually became the main source of energy. The invention of synthetic food led to a great increase in the world's population, and after the federation of the world it settled down at about twelve thousand million. As tide engines were developed, an ever-increasing use was made of their power; and before the human race had been in existence for a million years, the tide-power utilized aggregated a million million horsepower. The braking action of the tides was increased fiftyfold, and the day began to lengthen appreciably.

At its natural rate of slowing fifty thousand million years would have elapsed before the day became as long as the month, but it was characteristic of the dwellers on Earth that they never looked more than a million years ahead, and the amount of energy available was ridiculously squandered. By the year five million the human race had reached equilibrium; it was perfectly adjusted to its environment, the life of the individual was about three thousand years; and the individuals were 'happy', that is to say, they lived in accordance with instincts which were gratified. The tidal energy available was now fifty million million horsepower. Large parts of the planet were artificially heated. The continents were remodelled, but human effort was chiefly devoted to the development of personal relationships and to art and music, that is to say, the production of objects, sounds, and patterns of events gratifying to the individual.

Human evolution had ceased. Natural selection had been abolished, and the slow changes due to other causes were traced to their sources and prevented before very great effects had been produced. It is true that some organs found in primitive man, such as the teeth (hard, bone-like structures in the mouth), had disappeared. But largely on aesthetic grounds the human form was not allowed to vary greatly. The instinctive and traditional preferences of the individual, which were still allowed to influence mating, caused a certain standard body form to be preserved. The almost complete abolition of the pain sense which was carried out before the year five million was the most striking piece of artificial evolution accomplished. For us, who do not regard the individual as an end in itself, the value of this step is questionable.

Scientific discovery was largely a thing of the past, and men of a scientific bent devoted themselves to the more intricate problems of mathematics, organic chemistry, or the biology of animals and plants, with little or no regard for practical results. Science and art were blended in the practice of horticulture, and the effort expended on the evolution of beautiful flowers would have served to alter the human race profoundly. But evolution is a process more pleasant to direct than to undergo.

By the year eight million the length of the day had doubled, the Moon's distance had increased by 20 per cent, and the month was a third longer than it had been when first measured. It was realized that the Earth's rotation would now diminish rapidly, and a few men began to look ahead, and to suggest the colonization of other planets. The older expeditions had all been failures. The projectiles sent out from the Earth had mostly been destroyed by air friction, or by meteorites in interstellar space, and those which had reached the Moon intact had generally been smashed by their impact on landing. Two expeditions had landed there with oxygen supplies, successfully mapped the face of it which is turned away from the Earth, and signalled their results back. But return was impossible, and their members had died on the Moon. The projectiles used in the earlier expeditions were metal cylinders ten metres or less in diameter and fifty or more in length. They were dispatched from vertical metal tubes several kilometres in length, of which the lower part was embedded below the Earth, while the upper projected. In order to avoid atmospheric resistance these tubes were generally built in high mountains, so that when the projectile emerged it had relatively little air to go through. The air in the tube itself was evacuated and a lid on the top removed as the projectile arrived. It was started off by a series of mild explosions which served to give it a muzzle velocity of about five kilometres per second without causing too great a shock. When it had left the lower atmosphere it progressed on the rocket principle, being impelled forward by the explosion of charges in its tail. The empty sections of the tail were also blown backward as required. It could be turned from inside by rotating a motor, or by the crew walking round.

On arriving in the gravitational field of another planet its fall could be slowed by the discharge downward of more of its explosive cargo, and to check the final part of its fall various types of resistance were employed, and collapsible metal rods were extruded to break the shock of landing. Nevertheless, landing was generally fatal. As is well known, different principles are now employed. In particular, on leaving the atmosphere, wings of metallic foil of a square kilometre or more in area are spread out to catch the Sun's radiation pressure, and voyages are thus made on principles analogous to those employed in the ancient sailing-ships.

The desire for individual happiness, and the fact that it was achieved on Earth, made membership of such expeditions unpopular. The volunteers, who were practically committing suicide, were almost all persons whose mates had died prematurely, or whose psychology was for some reason so abnormal as to render them incapable of happiness. An expedition reached Mars successfully in the year 9,723,841, but reported that colonization was impracticable. The species dominant on that planet, which conduct its irrigation, are blind to those radiations which we perceive as light, and probably unaware of the existence of other planets; but they appear to possess senses unlike our own, and were able to annihilate this expedition and the only other which reached Mars successfully.

Half a million years later the first successful landing was effected on Venus, but its members ultimately perished owing to the unfavourable temperature conditions and the shortage of oxygen in its atmosphere. After this such expeditions became rarer.

In the year 17,846,151 the tide machines had done the first half of their destructive work. The day and the month were now of the same length. For millions of centuries the Moon had always turned the same face to the Earth, and now the Earth-dwellers could only see the Moon from one of their hemispheres. It hung permanently in the sky above the remains of the old continent of America. The day now lasted for forty-eight of the old days, so that there were only seven and a half days in the year. As the day lengthened the climate altered enormously. The long nights were intensely cold, and the cold was generally balanced by high temperatures during the day. But there were exceptions.

Mankind had appeared on Earth during a period characterized by high mountains and recurrent ice-ages. Mountain-building had indeed almost ceased, though some ranges and many volcanoes appeared during man's early life. But four ice-ages occurred shortly before history began, and a fifth had devastated parts of the northern continents during the second hundred thousand years of history. The ice had, however, been kept within relatively narrow limits by human endeavour. After the end of this period a huge co-operative effort of the human species had destroyed the remaining ice-fields. About the year 220,000 the ice-cap of Greenland had been gradually melted by the application of tidal energy, and soon after this the Arctic Ocean had become permanently ice-free. Later the Antarctic continent had been similarly treated. Through most of the first half of human history there was therefore no permanent ice or snow save on a few mountains. The climate throughout the Earth became relatively mild and uniform, as it had been through most of the time recorded by geology.

But as the Earth's rotation slowed down, its equator contracted, causing earthquakes and mountain-building on a large scale. A good deal of land emerged from the oceans, especially the central Pacific. And with the lengthening of the nights snow began to be deposited on the uplands in fairly large amounts; near the poles the Sun occasionally failed to melt it during the day, and even where it was melted the subsoil was often permanently frozen. In spite of considerable efforts, ice-fields and giant glaciers had already appeared when the Moon ceased to rise and set. Above them permanent anticyclones once more produced storms in the temperate regions, and rainless deserts in the tropics.

The animals and plants only partially adapted themselves to the huge fluctuations of temperature. Practically all the undomesticated mammals, birds, and reptiles became extinct. Many of the smaller plants went through their whole life cycle in a day, surviving only as seeds during the night. But most of the trees became extinct except when kept warm artificially.

The human race somewhat diminished in numbers, but there was still an immense demand for power for heating and cooling purposes. The tides raised by the Sun, although they only occurred fifteen times per year, were used for these ends, and the day was thus still further lengthened.

The Moon now began once more to move relative to the Earth, but in the opposite direction, rising in the west and setting in the east. Very gradually at first, but then with ever-increasing speed, it began to approach the Earth again, and appear larger. By the year 25,000,000 it had returned to the distance at which it was when man had first evolved, and it was realized that its end, and possibly the Earth's, was only a few million years ahead. But the vast majority of mankind contemplated the death of their species with less aversion than their own, and no effective measures were taken to forestall the approaching doom.

For the human race on Earth was never greatly influenced by an envisaged future. After physiology was discovered primitive men long continued to eat and drink substances which they knew would shorten and spoil their lives. Mineral fuels were also oxidized without much forethought. The less pigmented of the primitive races exhausted the fuel under the continents on which they lived with such speed that for some centuries the planet was dominated by the yellow variety resident in eastern Asia, where mining had developed more slowly; until they too had exhausted their fuel resources. The unpigmented men appear to have foreseen this event, but did little or nothing to prevent it, even when it was clearly only a few generations ahead. Yet they had before them the history of an island in the North Atlantic on which Newton and Darwin are said to have lived, and whose inhabitants were the first to extract mineral fuel and the first to exhaust it, after which they disappeared from the stage of history, although at one time they had controlled large portions of the Earth's land surface.

On the contrary, the Earth's inhabitants were often influenced in a curious way by events in the past. The early religions all attached great significance to such occurrences. If our own minds dwell more readily on the future, it is due largely to education and daily propaganda, but partly to the presence in our nuclei of genes such as H 149 and P 783 c, which determine certain features of cerebral organization that had no analogy on Earth. For this reason we have undertaken the immense labour necessary to tap the central heat of our planet, rather than diminish its rotation. Even now this process involves a certain annual loss of life, and this was very much greater at first, so much so as to forbid its imitation on the Earth, whose inhabitants generally valued their own lives and one another's.

But if most men failed to look ahead, a minority felt otherwise, and expeditions to Venus became commoner. After 284 consecutive failures a landing was established, and before its members died they were able to furnish the first really precise reports as to conditions on that planet. Owing to the opaque character of our atmosphere, the light signals of the earlier expeditions had been difficult to pick up. Infrared radiation which can penetrate our clouds was now employed.

A few hundred thousand of the human race, from some of whom we are descended, determined that though men died, man should live for ever. It was only possible for humanity to establish itself on Venus if it were able to withstand the heat and want of oxygen there prevailing, and this could only be done by a deliberate evolution in that direction first accomplished on Earth. Enough was known of the causes responsible for evolution to render the experiment possible. The human material was selected in each generation. All who were not willing were able to resign from participation, and among those whose descendants were destined for the conquest of Venus a tradition and an inheritable psychological disposition grew up such as had not been known on Earth for twenty-five million years. The psychological types which had been common among the saints and soldiers of early history were revived. Confronted once more with an ideal as high as that of religion, but more rational, a task as concrete as and infinitely greater than that of the patriot, man became once more capable of self-transcendence. Those members of mankind who were once more evolving were not happy. They were out of harmony with their surroundings. Disease and crime reappeared among them. For disease is only a failure of bodily function to adjust itself to the environment, and crime a similar failure in behaviour. But disease and crime, as much as heroism and martyrdom, are part of the price which must be paid for evolution. The price is paid by the individual, and the gain is to the race. Among ourselves an individual may not consider his own interests a dozen times in his life. To our ancestors, fresh from the pursuit of individual happiness, the price must often have seemed too great, and in every generation many who have now left no descendants refused to pay it.

The modes of behaviour which our ancestors gradually overcame, and which only recur as the rarest aberrations among ourselves, included not only such self-regarding sentiments as pride and a personal preference concerning mating. They embraced emotions such as pity (an unpleasant feeling aroused by the suffering of other individuals). In a life completely dedicated to membership of a super-organism the one is as superfluous as the other, though altruism found its place in the emotional basis of the far looser type of society prevalent on Earth.

In the course of ten thousand years a race had been evolved capable of life at one-tenth of the oxygen pressure prevalent on Earth, and the body temperature had been raised by six degrees. The rise to a still higher temperature, correlated as it was with profound chemical and structural changes in the body, was a much slower process. Projectiles of a far larger size were dispatched to Venus. Of 1734, only 11 made satisfactory landings. The crews of the first two of these ultimately perished; those of the next eight were our ancestors. The organisms found on Venus were built of molecules which were mostly mirror images of those found in terrestrial bodies. Except as sources of fat they were therefore useless for food, and some of them were a serious menace. The third projectile to arrive included bacteria which had been synthesized on Earth to attack l-glucose and certain other components of the organisms on Venus. Ten thousand years of laboratory work had gone to their making. With their aid the previous life on that planet was destroyed, and it became available for the use of man and the sixty terrestrial species which he had brought with him.

The history of our planet need not be given here. After the immense efforts of the first colonizers, we have settled down as members of a super-organism with no limits to its possible progress. The evolution of the individual has been brought under complete social control, and besides enormously enhanced intellectual powers we possess two new senses. The one enables us to apprehend radiation of wavelengths between 100 and 1,200 metres, and thus places every individual at all moments of life, both asleep and awake, under the influence of the voice of the community. It is difficult to see how else we could have achieved as complete a solidarity as has been possible. We can never close our consciousness to those wavelengths on which we are told of our nature as components of a super-organism or deity, possibly the only one in space-time, and of its past, present, and future. It appears that on Earth the psychological equivalent of what is transmitted on these wavelengths included the higher forms of art, music, and literature, the individual moral consciousness, and, in the early days of mankind, religion and patriotism. The other wavelengths inform us of matters which are not the concern of all at all times, and we can shut them out if we so desire. Their function is not essentially different from that of instrumental radio-communication on Earth. The new magnetic sense is of less importance, but is of value in flying and otherwise in view of the very opaque character of our atmosphere. It would have been almost superfluous on Earth. We have also recovered the pain sense, which had become vestigial on Earth, but is of value for the survival of the individual under adverse circumstances, and hence to the race. So rapid was our evolution that the crew of the last projectile to reach Venus were incapable of fertile unions with our inhabitants, and they were therefore used for experimental purposes.

During the last few million years the Moon approached the Earth rather rapidly. When it became clear that the final catastrophe could not be long delayed the use of tide-power was largely discontinued, according to the signals which reached us from the Earth, and wind and other sources of power were substituted. But the Earth-dwellers were sceptical as to whether the approaching rupture of the Moon would entail their destruction, and the spin of the

Earth–Moon system was still used to some extent as a source of power. In the year 36,000,000 the Moon was at only a fifth of its distance from the Earth when history had begun. It appeared twenty-five times as large as the Sun, and raised the sea-level by some 200 metres about four times a year. The effects of the tidal strain raised in it by the Earth began to tell. Giant landslips were observed in the lunar mountains, and cracks occasionally opened in its surface. Earthquakes also became rather frequent on the Earth.

Finally the Moon began to disintegrate. It was so near to the Earth as to cover about a twentieth of the visible heavens when the first fragments of rock actually left its surface. The portion nearest to the Earth, already extensively cracked, began to fly away in the form of meteorites up to a kilometre in diameter, which revolved round the Earth in independent orbits. For about a thousand years this process continued gradually, and finally ceased to arouse interest on the Earth. The end came quite suddenly. It was watched from Venus, but the earlier stages were also signalled from the Earth. The depression in the Moon's surface facing the Earth suddenly opened and emitted a torrent of white-hot lava. As the Moon passed round the Earth it raised the temperature in the tropics to such an extent that rivers and lakes were dried up and vegetation destroyed.

The colour changes on Earth due to the flowering of the plants which were grown on it for the pleasure of the human race, and which were quite visible from our planet, no longer occurred. Dense clouds were formed and gave some protection to the Earth. But above them the sea of flame on the Moon increased in magnitude, and erupted in immense filaments under the Earth's gravitation. Within three days the satellite had broken up into a ring of white-hot lava and dust. The last message received from the Earth stated that the entire human race had retired underground, except on the Antarctic continent, where however the ice-cap had already melted and the air temperature was 35 °C. Within a day from the Moon's breakup the first large fragment of it had fallen on the Earth. The particles formed from it were continually jostling, and many more were subsequently driven down. Through the clouds of steam and volcanic smoke which shrouded the Earth our astronomers could see but little, but later on it became clear that its tropical regions had been buried many kilometres deep under lunar fragments, and the remainder, though some traces of the former continents remain, had been submerged in the boiling ocean. It is not considered possible that any vestige of human life remains, nor can our spectroscopes detect any absorption bands of chlorophyll which would indicate the survival of plants.

The majority of the lunar matter has formed a ring round the Earth, like those of Saturn, but far denser. It is not yet in equilibrium, and fragments will continue to fall on the Earth for about another thirty-five thousand years. At the end of that period the Earth, which now possesses a belt of enormous mountains in its tropical regions, separated from the poles by two rings of sea, will be ready for recolonization. Preparations are being made for this event. We have largely sorted out the useful elements in the outer five kilometres or so of our planet, and it is proposed, when the Earth is reoccupied, to erect artificial mountains on both planets which will extend above the Heaviside layer and enable continuous radio-communication instead of light signals to be used between the two.

The old human race successfully cultivated individual happiness and has been destroyed by fire from heaven. This is not a cause for great regret, since happiness does not summate. The happiness of ten million individuals is not a millionfold the happiness of ten. But the unanimous co-operation of ten million individuals is something beyond their individual behaviour. It is the life of a super-organism.

If, as many of the Earth-dwellers hoped, the Moon had broken up quietly, their species might have lasted a thousand million years instead of thirty-nine million, but their achievement would have been no greater.

From the Earth it is proposed to colonize Jupiter. It is not certain that the attempt will succeed, for the surface temperature of that planet is 130 °C, gravitation is three times as intense as that on Venus, and over twice that on Earth, while the atmosphere contains appreciable quantities of thoron, a radioactive gas. The intense gravitation would of course destroy bodies as large as our own, but life on Jupiter will be possible for organisms built on a much smaller scale. A dwarf form of the human race about a tenth of our height, and with short stumpy legs but very thick bones, is therefore being bred. Their internal organs will also be very solidly built. They are selected by spinning them round in centrifuges which supply an artificial gravitational field, and destroy the less suitable members of each generation. Adaptation to such intense cold as that on Jupiter is impracticable, but it is proposed to send projectiles of a kilometre in length, which will contain sufficient stores of energy to last their inhabitants for some centuries, during which they may be able to develop the sources available on that planet. It is hoped that as many as one in a thousand of these projectiles may arrive safely. If Jupiter is successfully occupied the outer planets will then be attempted.

About 250 million years hence our solar system will pass into a region of space in which stars are far denser than in our present neighbourhood. Although not more than one in ten thousand is likely to possess planets suitable for colonization, it is considered possible that we may pass near enough to one so equipped to allow an attempt at landing. If by that time the entire matter of the planets of our system is under conscious control, the attempt will stand some chance of success. Whereas the best time between the Earth and Venus was one-tenth of a terrestrial year, the time taken to reach another stellar system would be measured in hundreds or thousands of years, and only a very few projectiles per million would arrive safely. But in such a case waste of life is as inevitable as in the seeding of a plant or the discharge of spermatozoa or pollen. Moreover, it is possible that under the conditions of life in the outer planets the human brain may alter in such a way as to open up possibilities inconceivable to our own minds. Our galaxy has a probable life of at least eighty million million years. Before that time has elapsed it is our ideal that all the matter in it available for life should be within the power of the heirs of the species whose original home has just been destroyed. If that ideal is even approximately fulfilled, the end of the world which we have just witnessed was an episode of entirely negligible importance. And there are other galaxies.'

EPILOGUE

There are certain criteria which every attempt, however fantastic, to forecast the future should satisfy. In the first place, the future will not be as we should wish it. The Pilgrim Fathers were much happier in England under King James I than they would be in America under President Coolidge. Most of the great ideals of any given age are ignored by the men of later periods. They only interest posterity in so far as they have been embodied in art or literature. I have pictured a human race on the Earth absorbed in the pursuit of individual happiness; on Venus mere components of a monstrous ant-heap. My own ideal is naturally somewhere in between, and so is that of almost every other human being alive today. But I see no reason why my ideals should be realized. In the language of religion, God's ways are not our ways; in that of science, human ideals are the products of natural processes which do not conform to them.

Secondly, we must use a proper time-scale. The Earth has lasted between one and eight thousand million years. Recorded human history is a matter of about six thousand. This period bears the same ratio to the Earth's life as does a space of two or three days to the whole of human history. I have no doubt that in reality the future will be vastly more surprising than anything I can imagine. But when we once realize the periods of time which our thought can and should envisage we shall come to see that the use, however haltingly, of our imaginations upon the possibilities of the future is a valuable spiritual exercise.

For one of the essential elements of religion is an emotional attitude towards the universe as a whole. As we come to realize the tiny scale, both temporal and spatial, of the older mythologies, and the unimaginable vastness of the possibilities of time and space, we must attempt to conjecture what purposes may be developed in the universe that we are beginning to apprehend. Our private, national, and even international aims are restricted to a time measured in human life-spans.

> And yonder all before us lie
> Deserts of vast eternity.

If it is true, as the higher religions teach, that the individual can only achieve a good life by conforming to a plan greater than his own, it is our duty to realize the possible magnitude of such a plan, whether it be God's or man's. Only so can we come to see that most good actions merely serve to stave off the constant inroads of chaos on the human race. They are necessary, but not sufficient. They cannot be regarded as active co-operation in the Plan. The man who creates a new idea, whether expressed in language, art, or invention, may at least be co-operating actively. The average man cannot do this, but he must learn that the highest of his duties is to assist those who are creating, and the worst of his sins to hinder them.

I do not see how anyone who has accepted the view of the universe presented by astronomy and geology can suppose that its main purpose is the preparation of a certain percentage of human souls for so much of perfection and happiness as is possible for them. This may be one of its purposes, but it can hardly be the most important. Events are taking place 'for other great and glorious ends' which we can only dimly conjecture. Professor Alexander, for example, in *Space, Time, and Deity*, suggests that the end towards which 'the whole creation groaneth and travaileth' is the emergence of a new kind of being which will bear the same relation to mind as do mind to life and life to matter. It is the urge towards this which finds its expression in the higher forms of religion. Without necessarily accepting such a view, one can express some of its implications in a myth. The numerical side of the myth is, I believe, correct, though whether tidal power could be utilized to the extent that I have suggested is a question for the engineers of the future.

Man's little world will end. The human mind can already envisage that end. If humanity can enlarge the scope of its will as it has enlarged the reach of its intellect, it will escape that end. If not, the judgement will have gone out against it, and man and all his works will perish eternally. Either the human race will prove that its destiny is in eternity and infinity, and that the value of the individual is negligible in comparison with that destiny, or the time will come

When the great markets by the sea shut fast
  All that calm Sunday that goes on and on;
When even lovers find their peace at last,
  And earth is but a star, that once had shone.

IS HISTORY A FRAUD?

EVERY generation must rewrite history. New facts become available, and old facts are interpreted anew. In the last century several new standpoints have been adopted, and in particular the attempt has been made to interpret history in terms of economics. But the greatest change has been in the extent of history. A hundred years ago it began about 700 BC. Before that time there were various legends. Those of the Bible were in a class apart, and they were treated as Sacred history, and put into a separate compartment from Profane, or ordinary, history. As long as one was compelled to believe in the literal truth of two mutually contradictory accounts of the great Mesopotamian flood, it was no use trying to disentangle the very considerable amount of historical fact embedded in these legends. And the effort of faith involved was relieved by a quite undue scepticism about other legendary events with a historical core, such as the siege of Troy and the story of the Minotaur.

The enlargement of our horizon began with the interpretation of Egyptian hieroglyphics. If Columbus doubled the field of geography by discovering America, Champollion in 1821 doubled the field of history by making possible the translation of documents some of which are over 4,000 years old. A generation later Rawlinson decoded the cuneiform script in which the languages of ancient Mesopotamia were inscribed on clay tablets. As a result of this, history now extends more than twice as far into the past as it did a century ago. It is true that the earliest date known with certainty is 2283 BC. At 11 a.m. on 8 March of that year occurred a total eclipse of the sun, which portended the sack of Ur by the Elamites. This ended the third dynasty of Ur, a city whose history at that time went back to before Noah's flood, which had not completely submerged it, though it laid down 6 feet of mud in its low-lying suburbs. The date of the flood is still doubtful, though probably somewhere between 4000 and 5000 BC. On the other hand, Woolley is quite confident as to the main sequence of events in southern Mesopotamia as far back as about 3500 BC, though the dates may well be a century out. Egyptian history appears to begin rather later than this.

Where there are no written or carved records it has been possible to construct a very rough picture of the more important events. Thus we have evidence, from sudden changes in the shapes of the skulls in graves and the objects found with them, of two prehistoric invasions of England. And still further back one discovers, though only in their dimmest outlines, a whole series of different stone ages, each with its characteristic skull-shape and art, until one reaches the half-men of Neanderthal, with great brow-ridges and no chins. They chipped flints in a crude way, and possessed fire, but, though they inhabited Europe for scores of thousands of years, they have not left a single work of art. Only by a perhaps misplaced courtesy do we call them men. It is against this background of barbarism that history stands out.

As history cannot exist in the absence of records, and as archaeology has already reached back to the origin of writing, from pictures, both in Egypt and Mesopotamia, it is unlikely that future research will ever extend our historical knowledge very much further into the past. We shall probably never know the name of any man, city, or nation, before 5000 BC. Most of historical research in the future will consist in the filling in of gaps. It is therefore possible today for the first time to take a bird's-eye view of history as a whole.

The picture so obtained proves, I think, that the history

taught today in our schools and universities is reliable in its details, but as a whole quite misleading. English history is taught as a progress in social organization, broken only by the decay of Roman civilization and its final overthrow by the Angles and Saxons. And the origins of our culture are traced back, on the one hand to the Greeks and Romans, who gradually built up a complete civilization with highly developed literature, art, and law, from rude beginnings; on the other to the Jews, who evolved most of the religious and ethical ideas which govern us today.

The truth is rather different. The curtain rises at Ur and other cities of the land then called Sumer in southern Mesopotamia, about 3500 BC, and reveals a fully developed civilization. They built well, using the arch, which only reached Europe 3,000 years later. They had cloth, wheeled vehicles, pottery, bronze, copper, silver and gold ware, a small amount of iron, sculpture, music, writing (on clay tablets), seals, and a complete social organization. And it is unfortunately quite clear to anyone who visited the British Museum in 1928 that their standard of taste in art was superior to our own today. They still killed servants to wait on dead princes in Kur-nu-gea (No Return Land), but this practice had been abandoned 500 years later. Though one cannot defend this custom, it is only fair to remember that in this enlightened age more people were killed in four years as a result of the death of the Archduke Franz Ferdinand than were sacrificed in the whole course of Sumerian history. When we get a clearer view of their civilization, about 2500 BC, we find sanitary conveniences with adequate drains in the houses, better than those of many English cottages today. There was a small standing army supported by a feudal system, with conscription in time of emergency for citizens. Slaves existed, but had some legal rights, and could own property. So could women, married or unmarried. There was a definite code of civil and criminal law, with professional judges. 4,500 years ago southern Mesopotamia was a great deal more civilized than is half the world today. Egypt was also civilized, though probably the average man or woman was worse off than in Mesopotamia. There was also a civilization in the valley of the Indus, of which we know very little, except that it must have been in contact, or have had a common origin, with that of Mesopotamia.

We do not yet know where civilization started. The Mesopotamians said that their ancestors came from the sea, that is the Persian Gulf. As they represented their gods as standing on mountains, it is conjectured that they came from a hilly country. Their culture cannot have come mainly from Egypt as is sometimes believed, unless some very serious mistakes indeed have been made with regard to dates.

At present the principal clue to the spot where civilization began comes from an entirely unexpected source, namely, plant genetics. Civilization is based, not only on men but on plants and animals. It needs a cultivated plant giving high yields of storable food, an animal to carry loads and pull carts or ploughs, and a plant or animal source of fibres. The principal plants available are the cereals, the soya bean, and the potato, and these are of very unequal value for biochemical reasons. For example, maize, as compared with wheat or oats, is very poor in vitamin B2. Hence populations living mainly on maize get a skin disease called pellagra. This is probably one reason why the maize-civilizations of central America never reached the level of the wheat, barley, and rice civilizations of the old world. The other reason is that America was very poor in domesticable animals. The buffalo is no substitute for the cow, and the llama a very poor one for the horse and sheep.

Hence, if it is possible to determine where cereals and cattle were first domesticated, we shall have gone a long way towards tracing civilization to its source. This task is being undertaken by Vavilov and other Russian scientists. Karl Mark's *Capital* has largely replaced the Bible in Russia today, and one of Marx's doctrines is that if we know how production is organized in a society we know the most important thing about it, and can even deduce its religious or philosophical system to a large extent. So Russian biologists are studying not only the domesticated animals and plants of today, but their ancestors which were the means of production in primitive societies. In the case of wheat the results are fairly clear. There are two distinct groups of wheat, which can only be hybridized with difficulty, and each can be traced to a definite centre. As that centre is approached, more and more different kinds of wheat are found, and these show all kinds of characters, such as purple shoots, which have been lost in the most cultivated varieties, and which are shown by breeding tests to be almost certainly primitive characters. One of these centres is in Abyssinia, the other, from which the more important group of wheats is derived, in or near south-eastern Afghanistan. The former is taken to be the original home of the agriculture that led up to Egyptian civilization, the latter the source of Indian and Mesopotamian wheats, and of the more important varieties grown in Europe and North America today. What is more, a great many other cultivated plants seem to have originated in one or the other of these centres. For example, rye, carrots, turnips, and some types of beans, lentils, flax, and cotton, seem to be of Afghan origin. At present the archaeology of these regions is quite untouched, but the results of excavation, especially in the Afghan area, are likely to be of extreme interest. Agriculture, if Vavilov is right, started in mountains, and only later spread to river valleys.

In the same way a knowledge of the origin of the dog would throw an immense amount of light on prehistory.

Dogs have been domesticated since neolithic times at least, probably for far longer than cattle, which is doubtless one reason why they fit better into human society. However, no one has yet any serious idea where they were first domesticated.

But to return to better-ascertained facts. Between about 3000 BC and AD 1400 there was very little improvement in the quality of civilization at its best. Yet it did spread out from its original centres in the valleys of the Nile, Euphrates, and Indus, to cover an ever wider area. This area sometimes contracted, as when our ancestors overran the western Roman Empire, when the Turks destroyed the civilization of Mesopotamia after a continuous run of over 4000 years, or when large areas of Central Asia dried up into deserts. It is probable that an important part in shifting the centres of civilization to more temperate countries was played by the malaria parasite and the hookworm *Ankylostoma*, which causes anaemia. These can only flourish in warm damp countries, and there is a certain amount of evidence that they have been spread about the world during the last 4,000 years.

Between 3000 BC and AD 1400 there were probably only four really important inventions, namely the general use of iron, paved roads, voting, and religious intolerance. Perhaps I should have added coinage and long-distance water-supply. Gunpowder had been known for a long time before AD 1400 in China, but did not begin to win battles in Europe till the seventeenth century. Somewhat before that date, however, it had helped to accelerate the decay of feudalism by diminishing the military value of castles. Knowledge progressed slowly, and we now know that we have greatly overestimated the originality of Greek mathematics. Babylonian mathematical astronomy was very highly advanced. Kidinnu, their last great astronomer, who lived about 400 BC, was a great deal more accurate in the numbers which he used in predicting eclipses and the like than any of his successors until about fifty years ago. His knowledge had, however, been forgotten in the interval, and his calculations were translated just too late to be of any serious value to astronomers. In Assyria the average educated man knew the multiplication table. As King Ashurbanipal put it in his autobiography, 'I recited the complicated multiplications and divisions which are not immediately apparent.' The same level was not reached in England till the late seventeenth century. Pepys was grown up when he learned his multiplication table.

As regards law, the code of King Dungi, who reigned in Ur about 2340 BC, compares quite favourably with that of King George IV of England a century ago. King Dungi's subjects kept slaves, though the slaves were allowed private property. They did not, however, hang children for theft. Their wives, unlike those of our great-grandfathers, were allowed private property, and if their husband took a concubine, instead of having no legal remedy at all, like English women up to 1923, they had the right to make their supplanter wash their feet and carry their chair to church, though she had also certain definite rights as against the husband. As the legal code gives a rough reflection of the moral standards of those who framed it, we may suppose that, on the whole, morals have not greatly improved during the course of history.

Christianity and other religions have, of course, on occasion been great weapons in the hands of moral reformers, but they have also been effectively used for the opposite purpose. To take an obvious example, slavery, and what is worse, slave-raiding, still exist in Christian Abyssinia, the latter evil nowhere else. And when Lloyd Garrison opened his anti-slavery campaign in Boston in 1830 he met with such opposition from all religious bodies that he was compelled to start in an infidel hall.

These facts must be weighed against the religious motives which prompted Wilberforce and Clarkson in their campaign against slavery in the British Empire. The balance is equally even in the case of other moral reforms.

Iron of a sort was known from a very early age, but it was only produced on a very large scale and of a useful type in the second millennium BC. At the siege of Troy, about 1200 BC, it was still an expensive novelty. It made a somewhat higher material level of civilization possible, but it also made war more efficient and terrible. Paved roads increased the possible size of the state, and voting made various republican forms of government possible, though democracy was extremely rare. The so-called democracies of the ancient world were almost invariably governments by associations of slave-owners. Religious intolerance (which was possibly invented by the Jews, and independently by the Zoroastrian Persians) had important effects in producing uniformity of culture, and was a great means of spreading civilization. The ancient Romans, who were not intolerant, could not conquer the Germans, and did not try to make them substitute Jupiter for Thor. Indeed they thought the two were the same. (I always have to remember this fact before I translate '*Jeudi*' into English, my natural tendency being to equate Jove with Woden.) St Boniface and other missionaries persuaded many of the Germans to leave Thor for Christ, and incidentally to adopt various Roman customs which went along with Christianity, just as modern missionaries diffuse trousers and football along with the gospel. In this way the Germans were ultimately civilized. But religious intolerance, both Christian and Muhammadan, also played a great part in lowering the level of civilization throughout what had been the Roman Empire.

Up till about AD 1400, then, civilization spread a great deal, but rose very little. It is only if we confine our attention to such areas as Western Europe, where it arrived very late, that it appears to have improved. In the fifteenth century a new process began. For thousands of years educated people had despised manual labour. This was natural enough when it was largely performed by slaves. But in the late Middle Ages things were different for three reasons. In the first place, the ruling military class were illiterate. Many kings could not sign their names. There was, however, a fair amount of education in other parts of the population. Secondly, thanks to St Benedict and certain other founders of religious orders, a large number of the clerical class, who were relatively educated, had a firsthand acquaintance with manual labour. Thirdly, the towns were very largely governed by the guilds, in which men who had become skilled workers rose to positions of wealth and power.

Hence the possibilities for experimental investigation on a large scale by educated men arose. The scientists of the past had investigated nature, but almost always by observation and not experiment, and they had never made elaborate apparatus. Plato had believed that the future of humanity lay in the hands of the philosopher who was also a king. He was wrong. The combination required was that of philosopher and craftsman. Modern physics began in Leiden, where the great Simon Stevinus founded statics in 1586 by a study of the principles underlying the lever and the sluice. Incidentally he invented decimals and influenced world history about as much as Napoleon or Washington by devising the system of defence of Holland by sluices. This enabled the Dutch to win the eighty years' war against the Spaniards, who were far better soldiers, and saved the Reformation. And modern industry began with printing about 1450. This invention was important not only because it cheapened books, but because it was the first example of mechanical mass production applied to articles formerly produced one by one.

Even so the old civilization might perhaps have been saved. The main principles which have guided scientific research ever since were laid down by Galileo, who first used the experimental method not merely as an occasional resort in difficult cases, but as a normal method of investigation. The man who is probably the greatest living experimentalist once said to me that but for Galileo and men like him he would never have thought of using experiment rather than unaided observation and thought to search out the nature of things. If Galileo and a few more like-minded men had been burned alive at an early age we might very possibly still be living under a civilization not greatly different from that of the Middle Ages.

But the progress of science was slow. Galileo died in 1642, and it was not till 161 years later that Symington's steam tug, *Charlotte Dundas*, towed two barges for 19½ miles on the Forth and Clyde Canal. Leeuwenhoek invented the first efficient microscope in about 1660, and it was two centuries before Pasteur used it to discover the cause of infectious diseases. It is only in the last hundred years that civilization, after six thousand years, has begun to change all through. But today the external conditions of life in civilized communities differ more from those of 1829 than did the conditions of 1829 from those at the time of Noah's flood. And this change, the real world revolution, has only just begun. We have gone an immense way in improving and organizing production and communication; we have nearly abolished water-borne and insect-borne diseases, and that is about all. Science has not yet been applied to most human activities. It can be, and I hope will be, applied to all.

The world is, of course, full of alleged applications of science outside the realms of production and hygiene, but the vast majority of them show no trace of scientific method. Thus there are numberless systems of education which are supposed to be based on scientific child psychology. But they are usually applied to small groups of children, in many cases to the children of unusually intelligent parents, brought up in unusually intelligent homes. If such children later turn out to be more successful than the average, this proves nothing at all. In order to prove the superiority of some new system, for example the Dalton plan, it will be necessary to follow up some thousands of average children educated under it, and some thousands educated on the ordinary system, and to find out which group on the average grow up into better citizens. This has not yet been done, and until it has been done it is ridiculous to talk about scientific method in education. Scientific method combines observation with experiment. Experiment without observation may be an enthralling occupation, but it is not science.

But the application of science to industry and medicine has entirely altered political problems. Until a few years ago every 'civilized' country really consisted of a small number of more or less civilized people among a multitude of uneducated poor who shared to a very slight extent in the benefits of civilization. Any equalization of incomes would merely have reduced the few to the level of the many, and destroyed what little culture existed. Socialism and civilization were obviously incompatible. Today the national income is large enough to admit of universal education, and it could be more evenly divided than it is at present without endangering science, art, or literature. That particular argument against socialism is no longer valid. And hygiene has provided another serious argument against our present economic system. We now live so long that a large proportion of the capital in many countries is in the hands of people over sixty years of age, who naturally show less enterprise than younger men and women. A good deal of socialism arises from irritation at this fact, though anti-socialists can fairly reply that a government official at forty commonly shows as little enterprise as an ordinary man at sixty-five.

For this reason history helps us very little in deciding for or against socialism. The situation of today is something entirely new. The old civilization, which had lasted for six thousand years, is in process of replacement by something which will differ from it as completely as it differed from savagery. History, as generally taught in schools, is the story of the political squabbles of the last two thousand years, and is, on the whole, rather a futile story. It becomes valuable when it is studied in detail, because it illustrates the psychology of politicians and that of crowds. Far more light is thrown on the English civil war by the fact that Charles I was afflicted with severe stammering in his youth than by the quaint legal arguments which he used to justify his ill-considered actions. This is why men and women today prefer to read biographies of historical characters rather than histories of the British Constitution. We have our Charles Is in politics today, and biographical history enables us to understand and pity them. But conventional history may lead us to share their delusion that they are now living in the eighteenth century, as Charles I apparently supposed that he was living in the Middle Ages.

The interpretation of history has tended to oscillate between two fallacies. The obvious fallacy is to regard it as the story of great men and great movements. But on a long view these very nearly cancel one another out.

The struggle between freedom and authority has gone on all through history, and any unbiased person must recognize that both parties at any moment have had a good deal of right on their side. And few of us can be whole-hearted partisans in any war of more than a hundred years past. In disgust with these great political figures we turn to the idealists who took no direct part in government, but produced novel ideas and points of view. Here, we like to think, are the real leaders of mankind.

> We are the music-makers and we are the dreamers of dreams,
> Wandering by lone sea-breakers, and sitting by desolate streams,
> World-losers and world-forsakers, on whom the pale moon gleams,
> But we are the movers and shakers of the world forever, it seems.
> We in the ages lying in the buried past of the earth
> Built Nineveh with our sighing, and Babel itself with our mirth,
> And o'erthrew them with prophesying to the old of the new world's worth,
> For each age is a dream that is dying, or one that is coming to birth.

I believe that this is as great a fallacy as the other. The dreamer of dreams can at most replace one set of symbolic ideas by another, the cross by the crescent, or the mother of the gods by the mother of God. After wars and revolutions, crusades and martyrdoms, the new dream is sometimes adopted. The world has been shaken, but there is very little evidence that it has been moved. But if the dreamers and the music-makers have not greatly altered the world by imposing their special dreams on it, the greatest of them have slightly raised the level of human life. We can meet the prospect of death with greater equanimity because Shakespeare wrote:

> Men must endure
> Their going hence, even as their coming hither:
> Ripeness is all.

We can love more passionately because Marvell told his coy mistress that

> The grave's a fine and private place,
> But none I think do there embrace.

And we can be better citizens of the universe, better botanists, even better horticulturists, because Jesus said, 'Consider the lilies how they grow; they toil not, they spin not, and yet I say unto you, that Solomon in all his glory was not arrayed like one of these.'

The reason for the relatively small ultimate effect of the dreamer is, I think, fairly clear. He or she is primarily concerned with the human spirit. Even today the workings of the spirit of man largely elude our intellectual grasp. In other words, psychology is not a science. Spiritual things must therefore be shown, if at all, in symbols, and these symbols are interpreted in different ways by different men, so that Blake could write:

> The vision of Christ which thou dost see
> Is my vision's chiefest enemy.
> Yours is the healer of mankind
> Mine speaks in parables to the blind.

Thus religion tends inevitably to crystallize into theology, and the letter to choke the spirit.

Ultimately I can see no reason to doubt that psychology will become scientific, with results of incalculable importance. Even today the first feeble attempts to introduce scientific method into it are producing a change in human thought and conduct only comparable with those which are generally brought about by a new religion.

Who, then, have been the real world-revolutionaries, the men who have done such deeds that human life after them could never be the same as before? I think that the vast majority of them have been skilled manual workers who thought about their jobs. The very greatest of them are perhaps two men or women whose real names will remain forever unknown, but whom we may call Prometheus and Triptolemus, the inventors of fire and agriculture. Prometheus, who was a Neanderthal man[1] with great brow-ridges and no chin, discovered how to keep a fire going, and how to use it to such advantage that his successors were induced to imitate his practice. Probably some later genius discovered how to kindle a fire by rubbing sticks together, and I like to imagine that it was a woman who first presented her astonished but delighted husband with a cooked meal. Fire was a very ancient invention, made in the early part of the old Stone Age, but apparently seeds were first systematically sown not so very long before the dawn of history. The immediate result was to make possible a fairly dense and settled population in which civilization was able to develop.

All through the historical period great inventions were made which were so clearly useful that they were bound to spread over the earth. Great intellectual discoveries were also made, but they were often forgotten because they led to no practical result. Thus the ancient Egyptian possessed a primitive kind of algebra. The chief algebraical papyrus known to us, which deals with simple equations, is called 'Directions for obtaining knowledge of all dark things.' But this algebra was forgotten and had to be reinvented, because it was not applied to any useful purpose, whereas the Egyptian methods of surveying have developed into those in use today. Today science is important because it is applied, and it is only the applicable portions of science which are reasonably sure of survival.

Compare the two greatest biologists of last century, Pasteur and Darwin. Pasteur's fundamental ideas are fairly sure of survival, because any nation that disbelieved in them would double its death-rate if it carried that disbelief into practice. But although Darwin's main ideas are accepted by

Recent excavations in China suggest that the ape-man *Sinanthropus* possessed fire. Prometheus lived longer ago than I thought.

most scientific men, no obvious disasters would follow their rejection.

Both in England and America there are religious bodies which are either anti-Pasteurian or anti-Darwinian. It is perfectly conceivable that during the next century the Roman Catholic Church may gain control of Europe, or the fundamentalists of North America. In either case, Darwinism will be proscribed, and the average man will not be much worse off on that account. But if in the next fifty years Darwin's ideas are applied to produce some great improvements in agriculture, hygiene, or politics, such a proscription will at once become more difficult. A government of consistent Christian Scientists, who refused to take preventive measures of a material kind against the spread of epidemic disease, would be far more dangerous than a government of fundamentalists. Darwin's intellectual achievement may have been as great as Pasteur's, but so far it has only led to a change in fashionable beliefs which may not be permanent, while Pasteur's has affected the whole structure of civilized society, and will probably go on doing so.

If I become Pope, which does not at present seem very probable, I shall at once take all the steps in my power to secure the canonization of Pasteur, who was, of course, a sincere Catholic. And I shall give the official blessing of the Church to some of the theories and practices which he introduced. But I shall point out the really weak points in Darwin's argument, which most defenders of the faith seem to miss completely, and anathematize them as errors.

It is significant that Pasteur was not only a great thinker but a superb technician, a man of immense manual skill who invented a great deal of the complex technique by which substances can be kept free from microbes, and one kind of microbe can be grown without contamination by others. Bacteriological theory is largely the verbalization of this technique. Pasteur clearly thought a great deal with his hands, Darwin rather little.

Many of the more historically important ideas were not at first put into words. They were technical inventions, which were at first handed down by imitation, and only slowly developed a verbal theory. When they did the theory was generally nonsense, but the practice sound. This was obviously the case, for example, until quite recently, with the extraction of metals from their ores. Certain methods worked, but no one knew why, and those who thought they knew were wrong. As the historical importance of production was not realized until recently, we shall never know who discovered iron-smelting, or, what would be more interesting, how he discovered it.

But there is another reason too. The first-rate technician is generally much more interested in his craft than in his personal fame, or even in his life. In order to obtain the necessary conditions to create a masterpiece or perfect a new process he is perfectly willing to lose himself in a glorious anonymity. The architects of many of the world's greatest buildings, like the great inventors, are often unknown, and generally mere names. The knowledge that this would be so would not have distressed them. Their attitude is summed up in one of the songs sung by airmen during the war.

> Take the cylinder out of my kidneys,
> The connecting rod out of my brain,
> The camshaft from under my backbone,
> And assemble the engine again.

The engine remains as their very real memorial. Similarly I am inclined to think that such men have been very largely responsible for so much of steady progress as is traceable behind the ebb and flow of history. The British Empire was made possible by the gradual improvement in navigation during the seventeenth and eighteenth centuries, and was consolidated by the steamship. The United States were united by railroads. The aeroplane is going to create the World State.

The point of view which I am urging is unpopular for two reasons, apart from the inevitable shortness of historical views until recently. In the first place, history is written by people impressed with the importance of their own political and religious views, and inevitably takes on the character of propaganda for them. But a more fundamental cause is as follows. Historians have inevitably thought in terms of words. They have read many books and documents. They have often been great stylists like Gibbon and Macaulay. They have realized the power of words to move multitudes. They have not been manual workers, and have seldom realized that man's hands are as important as and more specifically human than his mouth. Those intellectuals who have also been intelligent with their hands have mostly confined their writing to scientific and technical questions. Perhaps I ought to do so myself. But when I look at history, I see it as man's attempt to solve the practical problem of living. The men who did most to solve it were not those who thought about it, or talked about it, or impressed their contemporaries, but those who silently and efficiently got on with their work.

GOD-MAKERS

I AM fond of honorific titles, and I think that life has lost slightly in picturesqueness by their obsolescence. Besides his Majesty the King, his Holiness the Pope, and his Worship the Mayor, I should like to be able to speak of his Ferocity the Major-General, his Velocity the Air Marshal, and his Impiety the President of the Rationalist Press Association. Nevertheless, the most magnificent of all such titles belongs to a past which is not likely to be revived. It occurs in an inscription erected near Dratch in honour of the Roman emperors Diocletian and Maximian, who are described as 'Diis genitis et deorum creatoribus'—that is to say, 'Begotten by gods and creators of gods'. In those happy days the path to divinity was easier than in our irreligious age. A claim to divine descent might be made on somewhat slender grounds; but, as Diocletian and Maximian named thier successors, who, unless deposed during their lifetime, automatically became gods on dying, they could quite legitimately be claimed as god-makers.

It is only when we remember that they were first promulgated in an age of easy deification that we can properly assess the Christian dogmas of the divinity of Christ and the semi-divinity of Mary. At that time there was nothing in such assertions to surprise their pagan hearers, though unbelieving Jews might take a different view; and, as a God, Christ was clearly an improvement on Claudius or Hadrian. But if Christianity was probably the best of a number of competing creeds, it was also the product of an age when the moral and intellectual levels of the group of humanity round the Mediterranean were low—a fact sufficiently attested by their habit of indiscriminate god-making.

The saints, who perform so many of the minor functions of divinity in the Catholic scheme, are rather a mixed lot. Some men and women have achieved sanctity by virtue, others by hypocrisy, some again by sheer luck. Of this latter goodly fellowship none stand higher than St Protasus and St Gervaise. These worthy men (or possibly women, for, as we shall see, less is known about them than one might suppose) lived in northern Italy in late palaeolithic times, some ten to thirty thousand years ago, and died after doubtless unusually blameless lives. We do not know whether their beliefs on unascertainable matters were so coherent as to be dignified by the name of a religion. But they, or at least those who buried them, can hardly have believed that death was the end of Man's individual existence. For they took a great deal of trouble with corpses. First, the flesh was removed from the bones. They may have allowed it to decay, and have dug the skeleton up again after the lapse of some time. It is also possible that they stripped it from the bones soon after death. In this case it was probably eaten, at least in part, the meal being of a sacramental character, as still with some primitive peoples. If so, perhaps we must credit the eaters with religion of a kind, for the simple and materialistic belief that you can enter into communion with another person by eating him is at the basis of the most powerful religion of today.

The skeletons underwent a further treatment. Their heads were removed, and then the various bones were smeared with red ochre. We do not know the reasons for the first operation. Perhaps it was done to prevent the ghosts of the dead from walking. The meaning of the second is more obvious. The blood, as Holy Writ informs us, is the life. So, for a future life future blood is necessary. Ochre is a very good substitute for blood. It is red, and, not being susceptible of decay, may serve as a respiratory pigment during an eternal life. Moreover, recent biochemical research has demonstrated its peculiar suitability as a catalyst for those oxidations which are perhaps even more important in the future life than the present one. For spirit means breath, and the essential function of breathing is to supply oxygen.

Like the owners of other skeletons similarly fortified with red ochre (and many such have been found round Milan), the souls of Gervaise and Protasus, we may believe, chased the aurochs and the wild horse across the happy hunting-grounds, and tracked the woolly rhiconoceros to his lair in the Elysian swamps. But faith can work miracles, even on a woolly rhinoceros. Just as it can turn water into wine, and wine into blood (in spite of the fact that oenin, the pigment of grapeskins, unlike chlorophyll, that of leaves, stands in no chemical relationship to haemoglobin), so it can convert a woolly rhinoceros into a dragon. For in the town hall at Klagenfurt, in Carinthia, stands, or stood till recently, the skull of a woolly rhinoceros. To be more precise, the infidel palaeontologist would assign it to *Rhinoceros tichorinus*; but the noble knight who slew the dire monster in question said it was a dragon, and he ought to have known. Perhaps he really did kill the last survivor of this species. But more probably it had been extinct for some thousands of years, in which case it is not inconceivable that one of his villeins dug up the skull in his back garden.

Now, if the faith of a quite ordinary knight can transform a woolly rhinoceros into a dragon, why should not that of a particularly holy bishop convert two of its hunters into saints? At any rate, it did so. For a hundred centuries or more the spirits of Gervaise and Protasus hunted their ghostly quarry. But one day their pleasant, if monotonous, existence was sharply interrupted. Two angels appeared, and bore them away, perhaps slightly protesting, to the Christian heaven, where their spears were exchanged for harps and their skins for crowns. As they almost instantly began to work miracles in response to the prayers of the faithful, it appears that they must have adapted themselves to their new conditions more rapidly than might have been expected. Of course, several other cases have been recorded in which souls have gone to an apparently inappropriate heaven. Such were the souls of the penguins whose baptism by the myopic St Maël is reported by Anatole France, and that of the Christian knight Donander, which, as Cabell tells us in that most indecent, blasphemous, and amusing book *The Silver Stallion*, was transported to Valhalla by an unfortunate oversight, and subsequently elevated to Asgard on physiological grounds. And in our own days a respectable German medical officer of health has found himself in the Shinto heaven with Amaterasu and the divine Emperors. Robert Koch, the discoverer of the tubercle and cholera bacilli, and the joint founder with Pasteur of bacteriology, is worshipped as a god in at least one Japanese laboratory. It must at once be admitted that he appears to be quite an efficient god. Japan has produced a number of really excellent bacteriologists. But perhaps in another fifty years bacteriology may no longer be as important in medicine as it is now, and the divine Koch, like older gods, may prove a hindrance to medical progress by diverting effort into ineffective channels.

Spiritual events often have material causes, and we must now trace the mundane events which enriched heaven with its only palaeolithic saints. St Ambrose was one of the first batch of well-born Romans who, after its establishment as the State religion, entered the ministry of the Christian Church as a career. Like myself, he was unbaptized at the age of thirty-four; but, unlike me, he became a bishop before the application of that sacrament. He was not only a very able statesman, but a good poet—one of the pioneers of rhyming verse in Latin. In the year AD 385 he came into conflict with the secular authorities. The Dowager Empress Justina was an Arian, and demanded the use of a church in Milan for her co-religionists. The history of her conflict with Ambrose has been told by Gibbon in his twenty-seventh chapter. I shall not attempt to tell it again in detail.

Ambrose's tactics resembled those of Mr Gandhi today. While comparing the Empress to Jezebel and Herodias, he affected to deplore the rioting to which his language inevitably led. His methods were successful. The imperial court left Milan, and promulgated an edict of toleration for Arianism. The saint's protest against this tyrannical law led to a sentence of banishment. He blockaded himself in the cathedral with a pious bodyguard, including Augustine's mother, who kept up their spirits by singing his newly invented rhyming hymns, which brought frequent tears to the eyes of the future Saint Augustine, who had recently been baptized.

During the siege, in response to a vision, he dug for the bones of Gervaise and Protasus. They were found under a church floor, and it was revealed to St Ambrose that they had suffered martyrdom as Christians under Nero. The multitude were impressed not only by the miraculous freshness of the respiratory pigment of the martyrs, but by the large size of their bones. The Cro-Magnon race, to which the martyrs probably belonged, were, of course, very tall. The bones were carried with due pomp to the Ambrosian basilica. On the way a number of demons were expelled from lunatics, and a man called Severus, who had been blind for some years, was cured on touching the bier of the saints. St Augustine was in Milan at the time, and records these miracles, which were entirely successful in reinforcing the effect of the hymns. The soldiers did not dare to risk the bloodshed which would have been necessary to effect the capture of St Ambrose. Shortly afterwards the edict of toleration in favour of the Arians was withdrawn, and the illustrious examples of Gervaise and Protasus did much to confirm the general belief in the efficacy of relics. It only remains to add that twenty-four years after the discovery of the palaeolithic saints Rome was sacked by the Arian Goths. This time the Trinitarian saints were unable to rise to the occasion. Alaric was made of different stuff from Justina.

But saints are, after all, not gods; and a similar story, though involving a different red pigment, comes down to us from an age nearer to our own. At the Last Supper Jesus is reported to have said of the bread and wine: 'Take, eat; this is my body', and 'This is my blood of the new testament, which is shed for many.'

Personally, I am not one of those who find it probable that Jesus is a mainly mythical figure. A large number of his sayings seem to me to cohere as expressions of a definite and quite human character, which could hardly have been invented by disciples who wished to prove his divinity. He used figurative language about himself, calling himself, for example, the door and the vine. His self-identification with bread and wine is on a par with these utterances. But by a more or less fortuitous chain of events it has been taken much more seriously. One can imagine developments of Christianity in which every church door or every vine was identified with Jesus, as a pious Hindu may identify every cow with Agni. The actual form of the transubstantiation dogma appears to be due to three facts—the type of mystery religion flourishing in the early days of Christianity, the peculiarities of Latin and Greek grammar, and the activities of a particular god-making bacillus, which, besides upholding the views of the Angelic Doctor, St Thomas Aquinas, founded a college at each of our two older universities.

The importance of sacramental meals in mystery religions has been sufficiently stressed by others. If today we find it difficult to imagine how so much emotion could gather round the act of eating, we must remember that the majority of the early Christians were so poor as to have firsthand experience of real hunger. To most of them food must have presented itself not as a source of mildly pleasant sensations, but vividly as a life-giver.

Once Jesus had been identified with the sacramental meal, it was inevitable that some theory of that identity should be developed. The philosophers got busy. The only tools of philosophers, until very recently, were words, and the art of using words correctly was called 'logic'. In fact, words are well adapted for description and the arousing of emotion, but for many kinds of precise thought other symbols are much better. Russell and Whitehead were perhaps the first philosophers to take this fact seriously. But a perusal of their books makes it clear that even a greatly improved symbolism leaves room for a very comprehensive disagreement on fundamental tenets.

The European languages are characterized by a highly developed system of adjectives. For example, an Arab, instead of describing the Board of the RPA as infidel men, would call them fathers of infidelity; and I gather that a Chinese might also avoid the use of an adjective in a somewhat untranslatable manner. Now, the philosophy of the Middle Ages was the work of men who were ignorant of nature, but learned in Latin grammar. Neglecting the verbs, they tried to describe the universe in terms of substantives and adjectives, to which they attributed an independent existence under the names of substances and accidents or attributes. Modern physicists are engaged in a somewhat similar attempt to describe it in terms of verbs only, their favourite verb at the moment being to undulate, or wiggle. They are not concerned with what wiggles.

The scholastic philosophy, like any other, led to results calculated to alarm the pious. The soul was in danger of becoming a mere adjective of the body, and was therefore relegated to a special category of 'substantial forms', thus rendering it sufficiently durable to withstand eternal punishment. With such highly developed attributes, substance might have disappeared altogether had not a place been found for it by the genius of St Thomas Aquinas. St Thomas, it is said, was one of the fattest men who ever lived, and in his latter years could carry out the ritual of the mass only at a specially constructed concave altar. Hence his capacity for levitation was even more miraculous than that of lighter saints. In spite of the distance which separated him, in middle age, from the consecrated elements, he was able to observe that no perceptible change occurred when the bread and wine were converted into the body and blood of Christ. Very well, said he, in an excellent hymn, most inadequately rendered in the English hymn-book: 'Praestet fides supplementum sensuum defectui' (Let faith supplement the deficiency of the senses). It did. At the critical moment the substance of the bread and wine were converted into God; but, as all the accidents were unaltered, no perceptible difference occurred. Fortunately, he did not draw the full consequences from his theory. For, if no one could notice the difference when a piece of bread is converted into God, it would appear that the converse operation might also be imperceptible, and no one would notice any change if the object of St Thomas's worship were converted into a wafer or some other inanimate object. It is also interesting to note that, while St Thomas was a realist about things in general, he anticipated the views of Bishop Berkeley when it came to the consecrated elements. For he believed that their sensible qualities were directly caused and supported by the deity latent in them.

Now, the dogma of transubstantiation, which needed such strange intellectual props, was not merely based, like many theological dogmas, on traditions of past events which had been brooded over by successive generations of the pious. It was grounded on a series of very well-attested miracles. Not only had individual ecstatics seen visions of Jesus in the host, but large numbers of people had seen hosts bleeding. The first of such events which is known to me occurred in England about AD 900, in the presence of Archbishop Odo. Among the most famous is the miracle of Bolsena (also known as the miracle of the Bloody Corporal), which is portrayed in Raphael's well-known picture, and converted a priest who doubted transubstantiation. Allowing for a certain amount of exaggeration for the glory of God, I see no reason to disbelieve in these miracles. Their nature becomes very probable from the way in which they tended to occur in series, especially in Belgium. A 'bleeding host' appeared in a certain church. The faithful went to adore it, and fairly soon others appeared in the vicinity. There is very strong reason to suppose that we have to deal with an outbreak of infection of bread by *Bacillus prodigiosus* (the miraculous bacillus), which would naturally be spread by human contacts. Their organism grows readily on bread, and produces red patches, which the eye of faith might well take for blood.

The miracle of Bolsena appears to have finally converted Pope Urban IV to the views, not only of St Thomas, but of his contemporary, St Juliana of Liège, one of the two women who have initiated important changes in Catholic practice, the other being St Marie Marguerite Alacocque, the initiator of the cult of the Sacred Heart. St Juliana had a vision of the moon with a black spot on it, and was told that the moon signified the Church, the spot being the absence of a special cult of Christ's body. As a result of this vision the Bishop of Liège instituted the feast of Corpus Christi, and in 1264 Pope Urban IV, who had been Archdeacon of Liège, made its celebration compulsory throughout Western Europe. The office for the feast was written by St Thomas Aquinas. In honour of Christ's body a college was founded at Cambridge within the next century, though the corresponding establishment at Oxford dates back only to shortly before the Reformation. There is no record of what St Juliana said to the angel who told her about the activities of the poet Kit Marlowe, student of Corpus Christi College, Cambridge. For it appears from the record of his 'damnable opinion' that he was a remarkably militant rationalist, while a spy stated that he was 'able to shewe more sound reasons for Atheisme than any devine in Englande is able to geve to prove devinitie'. Perhaps, however, such things are kept from the ears of the blessed.

Unfortunately, *Bacillus prodigiosus* did not confine its efforts to inspiring queer metaphysics and founding colleges. If a bleeding host was God's body, any bit of bread which appeared to bleed was a host, presumably stolen and desecrated. Throughout the ages of faith the same incidents reoccurred. A piece of bread in a house started to 'bleed'. An informer, generally a servant, went to the authorities. The family were tortured, and finally confessed to having stolen or bought a consecrated wafer and run daggers through it. They were then generally burned alive. Such an incident was often a signal for a massacre of Jews, as in the pogrom of 1370, commemorated in the disgusting stained-glass windows of the cathedral of Ste Gudule at Brussels, and in the French outbreaks of 1290 and 1433. Sometimes the victims were gentiles, as in the case recorded by Paolo Uccello in a series of panels which were on view at the London exhibition of Italian painting in 1930. Doubtless among them were a few fools who were genuinely celebrating black masses; but the emphasis laid on the blood in contemporary accounts seems to incriminate *Bacillus prodigiosus*. In England the belief in transubstantiation ceased abruptly in the sixteenth century to be part of the law of the land. 'Hoc est corpus' became *hocus pocus*. But in France the attempt to make injuries to consecrated wafers a capital offence, as deicide, was one of the causes of the revolution of 1830.

So much for *Bacillus prodigiosus*, an organism which produced a delusion more serious than many diseases. But this god-making tendency seems to be one of the more unfortunate vices to which the human intellect is subject. We cannot observe a remarkable phenomenon without postulating something behind it. So far, so good; but we then proceed, if we are not careful, to endow that something with a personality, and deduce the oddest ethical implications—for example, that it is wrong to stick knives through certain pieces of bread. The same tendency operates in the sphere of science. A generalization is made from certain facts, and called a Law of Nature. This is then supposed to acquire, in some quite unexplained way, an ethical value, and to become a norm for conduct. Thus Darwin stated, probably quite correctly, that evolution had been mainly due to natural selection—i.e. the elimination of certain individuals, called the unfit, in each generation. The obvious comment was: 'So much the worse for nature; let us try to control our own evolution in some other way.' But a number of theorists, including even a few second-rate biologists, seem to have regarded it as an excuse for imitating nature. The weak, it was said, should be eliminated in various ways, and various forms of internecine struggle, from war to economic competition, were justified by an appeal to nature, which was only justifiable if nature represented God's unalterable plan—a view which these writers did not generally hold. The fact that in most civilized communities the poor breed more quickly than the rich shows that, from a Darwinian point of view, the poor are on the whole fit and the rich unfit. To call the rapidly breeding sections of the community unfit is certainly bad Darwinism. They may be undesirable, but that is another matter. To attempt to suppress them in the name of

Darwinism is an example of muddled thinking arising out of a partial deification of a law of nature.

Is the god-making tendency ineradicable, or may we hope that it will gradually die out or be sublimated into other channels? As long as it goes on there is very little chance for the development of a rational ethic based on the observable consequences of our actions. To answer this question one must consider the most important grounds for atheism. Perhaps the simplest hypothesis about the universe is that it has been designed by an almighty and intelligent creator. Darwin showed that much of the apparent design could be explained otherwise; but there still remains a group of facts, such as those collected by L. J. Henderson in *The Fitness of the Environment*, which are at present more readily conformable with the design theory than with any other. It is on the ethical side that theism has broken down most completely. For an almighty and all-knowing creator cannot also be all-good. It has only been possible to believe in all-powerful gods by attributing to them one or more of the seven deadly sins. The Graeco-Roman gods were at first conceived of as sharing all man's moral infirmities. Later, as their characters were idealized, their failure to improve matters here below was attributed to what was essentially sloth rather than active cruelty.

With Christianity, the Deity became more actively interested in human affairs, and it was necessary to attribute to him the darker defects of pride and wrath. His pride was particularly offended by the attempts of Satan and Adam to become like him, and his wrath visited the sin of the latter upon his descendants during thousands of years. A robust spirit like Thomas Paine could still see justice in the universe. It is to more delicate minds like that of Shelley that we look for the development of atheism on ethical grounds. The turning-point came perhaps when, under the influence of the utilitarians, the State set itself to be less cruel than nature or the hell-filling god of the clergy. We do not condemn our worst criminals to anything as bad as an inoperable cancer involving a nerve trunk. Dartmoor, our nearest equivalent to hell, has its alleviations, and, what is more, a hope of ultimate release. It became impossible to believe that the creator of the universe, even of a universe which did not include hell, was worthy of our moral admiration.

Christianity had, of course, attempted to meet such a criticism by the doctrine that God had become a man and suffered with men. This defence is based on the celebrated hypothesis that two blacks make a white, known to moralists as the retributive theory of punishment. The theory that a wrong act deserves the infliction of suffering is part of Christian ethics, and is responsible for any amount of cruelty even today. And the participation of God in human suffering, while admirable in a finite deity like Heracles, does not absolve an almighty power from the blame of having created suffering humanity.

Our present-day theists generally find two ways out of the dilemma. Either suffering is needed to perfect human character, or God is not almighty. The former theory is disproved by the fact that some people who have suffered very little, but have been fortunate in their ancestry and education, have very fine characters. The objection to the second is that it is only in connection with the universe as a whole that there is any intellectual gap to be filled by the postulation of a deity. And a creator could presumably create whatever he or it wanted. The evolution of life on earth can be pretty satisfactorily explained if we make certain assumptions about matter and life. The origin of the heavenly bodies presents greater difficulties, as will be apparent to any reader of Jeans's *The Universe Around Us*. The theory of creation is essentially a refusal to think back beyond a certain time in the past when it becomes difficult to follow the chain of causation. To hold such a belief is, therefore, always an excuse for intellectual laziness, and generally a sign of it. Probably we are waiting for a new Darwin to explain stellar evolution. But meanwhile an almighty deity would at least explain the apparent irreversibility of natural processes, while a finite deity struggling against the imperfections of matter would explain nothing whatever; and I know of no scientific facts which point to such a hypothesis. Humanity, or any other aggregate of such a kind, may very well take the place of god in an ethical system, but is not a god in any intelligible sense of that term.

Hence, so long as, on the one hand, scientific knowledge is preserved and expanded, and on the other man keeps his ethical standards above those of nature, the prospects for god-makers are by no means as rosy as they were in the past. I do not, however, think that the only alternatives to theism are agnosticism or any of the various forms of materialism, even though I should call myself an agnostic if forced to classify myself. There is a great deal of evidence that the universe as a whole possesses certain characters in common with the human mind. The materialist can agree with this statement, as he regards the mind as a special aspect of one small fraction of the universe in physical relation with the rest. The idealist regards our knowledge of mind as knowledge from inside, and therefore more satisfactory than our knowledge of matter. Unfortunately, there is a tendency to identify the absolute—i.e. the universe considered in its mind-like aspect—as in some sort an equivalent of God. I cannot see the cogency of this view. The absolute is not a creator, nor a soul animating otherwise inert matter, but just the universe looked at from the most comprehensive possible point of view. It cannot be identified with any of its constituents, though in the opinion of absolute idealists the human mind is more like it than is any other known finite existent.

Such a philosophy does, as a matter of fact, supply a fairly satisfactory emotional substitute for theism. It leads one to feel at home in the universe, and yet does not lend itself readily to the attribution of supernatural qualities to finite objects or finite events, which is the essence of all religions. Unfortunately, the history of Hinduism shows that it is compatible with religion in some of its least savoury forms. Brahma is the absolute; but, though he, or it, is venerated, he is not the centre of any important cult. Worship is reserved for Vishnu, Siva, and other minor gods and goddesses. For god-making has been carried out on a very large scale in India. But Brahma at least offers the philosophical Hindu an opportunity of 'turning his back on heaven', while preserving his piety—a gesture impossible to a European.

If this be taken as a condemnation of absolute idealism, it should be noted that in spiritualism we have the beginnings of a new religion, which can exist quite apart from any belief in a supreme deity, and often does so on the continent of Europe, though British and American spiritualists generally preserve a more or less Christian background. Clearly spiritualism demands scientific investigation, which would disclose remarkable facts—possibly of the type in which spiritualists believe—more probably concerning the psychology of small groups. As things are, the spiritualists are engaged in the same early stages of god-making as the primitive races, who are still mainly animists and ancestor worshippers. Unless the process is checked, spiritualism will presumably evolve into a fully developed religion, with sacred objects, intolerance, and that vast diversion of effort into fruitless channels which is in some ways the most characteristic feature of the religions.

I notice among many of my rationalist friends a lack of interest in the history of religions, which is quite natural when one has examined their fully developed forms and found them unsatisfactory. Nevertheless, the god-making tendency is always with us, and only by a study of its past are we likely to be able to curb its development in the present.

THE ORIGIN OF LIFE

UNTIL about 150 years ago it was generally believed that living beings were constantly arising out of dead matter. Maggots were supposed to be generated spontaneously in decaying meat. In 1668 Redi showed that this did not happen provided insects were carefully excluded. And in 1860 Pasteur extended the proof to the bacteria which he had shown were the cause of putrefaction. It seemed fairly clear that all the living beings known to us originate from other living beings. At the same time Darwin gave a new emotional interest to the problem. It had appeared unimportant that a few worms should originate from mud. But if man was descended from worms such spontaneous generation acquired a new significance. The origin of life on the earth would have been as casual an affair as the evolution of monkeys into man. Even if the latter stages of man's history were due to natural causes, pride clung to a supernatural, or at least surprising, mode of origin for his ultimate ancestors. So it was with a sigh of relief that a good many men, whom Darwin's arguments had convinced, accepted the conclusion of Pasteur that life can originate only from life. It was possible either to suppose that life had been supernaturally created on earth some millions of years ago, or that it had been brought to earth by a meteorite or by micro-organisms floating through interstellar space. But a large number, perhaps the majority, of biologists, believed, in spite of Pasteur, that at some time in the remote past life had originated on earth from dead matter as the result of natural processes.

The more ardent materialists tried to fill in the details of this process, but without complete success. Oddly enough, the few scientific men who professed idealism agreed with them. For if one can find evidences of mind (in religious terminology the finger of God) in the most ordinary events, even those which go on in the chemical laboratory, one can without much difficulty believe in the origin of life from such processes. Pasteur's work therefore appealed most strongly to those who desired to stress the contrast between mind and matter. For a variety of obscure historical reasons, the Christian Churches have taken this latter point of view. But it should never be forgotten that the early Christians held many views which are now regarded as materialistic. They believed in the resurrection of the body, not the immortality of the soul. St Paul seems to have attributed consciousness and will to the body. He used a phrase translated in the revised version as 'the mind of the flesh', and credited the flesh with a capacity for hatred, wrath, and other mental functions. Many modern physiologists hold similar beliefs. But, perhaps unfortunately for Christianity, the Church was captured by a group of very inferior Greek philosophers in the third and fourth centuries AD. Since that date views as to the relation between mind and body which St Paul, at least, did not hold, have been regarded as part of Christianity, and have retarded the progress of science.

It is hard to believe that any lapse of time will dim the glory of Pasteur's positive achievements. He published singularly few experimental results. It has even been suggested by a cynic that his entire work would not gain a doctorate of philosophy today! But every experiment was final. I have never heard of anyone who has repeated any experiment of Pasteur's with a result different from that of the master. Yet his deductions from these experiments were sometimes too sweeping. It is perhaps not quite irrelevant that he worked in his latter years with half a brain. His right cerebral hemisphere had been extensively wrecked by the bursting of an artery when he was only forty-five years old;

and the united brain-power of the microbiologists who succeeded him has barely compensated for that accident. Even during his lifetime some of the conclusions which he had drawn from his experimental work were disproved. He had said that alcoholic fermentation was impossible without life. Buchner obtained it with a cell-free and dead extract of yeast. And since his death the gap between life and matter has been greatly narrowed.

When Darwin deduced the animal origin of man a search began for a 'missing link' between ourselves and the apes. When Dubois found the bones of *Pithecanthropus* some comparative anatomists at once proclaimed that they were of animal origin, while others were equally convinced that they were parts of a human skeleton. It is now generally recognized that either party was right, according to the definition of humanity adopted. *Pithecanthropus* was a creature which might legitimately be described either as a man or an ape, and its existence showed that the distinction between the two was not absolute.

Now the recent study of ultramicroscopic beings has brought up at least one parallel case, that of the bacteriophage, discovered by d'Herelle, who had been to some extent anticipated by Twort. This is the cause of a disease, or, at any rate abnormality, of bacteria. Before the size of the atom was known there was no reason to doubt that

> Big fleas have little fleas
> Upon their backs to bite 'em;
> The little ones have lesser ones,
> And so ad infinitum.

But we now know that this is impossible. Roughly speaking, from the point of view of size, the bacillus is the flea's flea, the bacteriophage the bacillus's flea; but the bacteriophage's flea would be of the dimensions of an atom, and atoms do not behave like fleas. In other words, there are only about as many atoms in a cell as cells in a man. The link between living and dead matter is therefore somewhere between a cell and an atom.

D'Herelle found that certain cultures of bacteria began to swell up and burst until all had disappeared. If such cultures were passed through a filter fine enough to keep out all bacteria, the filtrate could infect fresh bacteria, and so on indefinitely. Though the infective agents cannot be seen with a microscope, they can be counted as follows. If an active filtrate containing bacteriophage be poured over a colony of bacteria on a jelly, the bacteria will all, or almost all, disappear. If it be diluted many thousand times, a few islands of living bacteria survive for some time. If it be diluted about ten million times, the bacteria are destroyed round only a few isolated spots, each representing a single particle of bacteriophage.

Since the bacteriophage multiplies, d'Herelle believes it to be a living organism. Bordet and others have taken an opposite view. It will survive heating and other insults which kill the large majority of organisms, and will multiply only in presence of living bacteria, though it can break up dead ones. Except perhaps in presence of bacteria, it does not use oxygen or display any other signs of life. Bordet and his school therefore regard it as a ferment which breaks up bacteria as our own digestive ferments break up our food, at the same time inducing the disintegrating bacteria to produce more of the same ferment. This is not as fantastic as it sounds, for most cells while dying liberate or activate ferments which digest themselves. But these ferments are certainly feeble when compared with the bacteriophage.

Clearly we are in doubt as to the proper criterion of life. D'Herelle says that the bacteriophage is alive, because, like the flea or the tiger, it can multiply indefinitely at the cost of living beings. His opponents say that it can multiply only as long as its food is alive, whereas the tiger certainly, and the flea probably, can live on dead products of life. They suggest that the bacteriophage is like a book or a work of art, which is constantly being copied by living beings, and is therefore only metaphorically alive, its real life being in its copiers.

The American geneticist Muller has, however, suggested an intermediate view. He compares the bacteriophage to a gene—that is to say, one of the units concerned in heredity. A fully coloured and a spotted dog differ because the latter has in each of its cells one or two of a certain gene, which we know is too small for the microscopist to see. Before a cell of a dog divides this gene divides also, so that each of the daughter-cells has one, two, or none according with the number in the parent cell. The ordinary spotted dog is healthy, but a gene common among German dogs causes a roan colour when one is present, while two make the dog nearly white, wall-eyed, and generally deaf, blind, or both. Most of such dogs die young, and the analogy to the bacteriophage is fairly close. The main difference between such a lethal gene, of which many are known, and the bacteriophage, is that the one is only known inside the cell, the other outside. In the present state of our ignorance we may regard the gene either as a tiny organism which can divide in the environment provided by the rest of the cell; or as a bit of machinery which the 'living' cell copies at each division. The truth is probably somewhere in between these two hypotheses.

Unless a living creature is a piece of dead matter plus a soul (a view which finds little support in modern biology) something of the following kind must be true. A simple organism must consist of parts A, B, C, D, and so on, each of which can multiply only in presence of all, or almost all, of the others. Among these parts are genes, and the bacteriophage is such a part which has got loose. This hypothesis becomes more plausible if we believe in the work of Hauduroy, who finds that the ultramicroscopic particles into which the bacteria have been broken up, and which pass through filters that can stop the bacteria, occasionally grow up again into bacteria after a lapse of several months. He brings evidence to show that such fragments of bacteria may cause disease, and d'Herelle and Peyre claim to have found the ultramicroscopic form of a common staphylococcus, along with bacteriophage, in cancers, and suspect that this combination may be the cause of that disease.

On this view the bacteriophage is a cog, as it were, in the wheel of a life cycle of many bacteria. The same bacteriophage can act on different species, and is thus, so to say, a spare part which can be fitted into a number of different machines, just as a human diabetic can remain in health when provided with insulin manufactured by a pig. A great many kinds of molecule have been got from cells, and many of them are very efficient when removed from it. One can separate from yeast one of the many tools which it uses in alcoholic fermentation, an enzyme called invertase, and this will break up six times its weight of cane-sugar per second for an indefinite time without wearing out. As it does not form alcohol from the sugar, but only a sticky mixture of other sugars, its use is permitted in the United States in the manufacture of confectionery and cake-icing. But such fragments do not reproduce themselves, though they take part in the assimilation of food by the living cell. No one supposes that they are alive. The bacteriophage is a step beyond the enzyme on the road to life, but it is perhaps an exaggeration to call it fully alive. At about the same stage on the road are the viruses which cause such diseases as small-pox, herpes, and hydrophobia. They can multiply only in living tissue, and pass through filters which stop bacteria.

With these facts in mind we may, I think, legitimately

speculate on the origin of life on this planet. Within a few thousand years from its origin it probably cooled down so far as to develop a fairly permanent solid crust. For a long time, however, this crust must have been above the boiling-point of water, which condensed only gradually. The primitive atmosphere probably contained little or no oxygen, for our present supply of that gas is only about enough to burn all the coal and other organic remains found below and on the earth's surface. On the other hand, almost all the carbon of these organic substances, and much of the carbon now combined in chalk, limestone, and dolomite, were in the atmosphere as carbon dioxide. Probably a good deal of the nitrogen now in the air was combined with metals as nitride in the earth's crust, so that ammonia[1] was constantly being formed by the action of water. The sun was perhaps slightly brighter than it is now, and as there was no oxygen in the atmosphere the chemically active ultraviolet rays from the sun were not, as they now are, mainly stopped by ozone (a modified form of oxygen) in the upper atmosphere, and by oxygen itself lower down. They penetrated to the surface of the land and sea, or at least to the clouds.

Now, when ultraviolet light acts on a mixture of water, carbon dioxide, and ammonia, a vast variety of organic substances are made, including sugars and apparently some of the materials from which proteins are built up. This fact has been demonstrated in the laboratory by Baly of Liverpool and his colleagues. In this present world such substances, if left about, decay—that is to say, they are destroyed by micro-organisms. But before the origin of life they must have accumulated till the primitive oceans reached the consistency of hot dilute soup. Today an organism must trust to luck, skill, or strength to obtain its

[1] Since the above was written, ammonia has been detected spectroscopically in the atmosphere of the outer planets by Wildt.

food. The first precursors of life found food available in considerable quantities, and had no competitors in the struggle for existence. As the primitive atmosphere contained little or no oxygen, they must have obtained the energy which they needed for growth by some other process than oxidation—in fact, by fermentation. For, as Pasteur put it, fermentation is life without oxygen. If this was so, we should expect that high organisms like ourselves would start life as anaerobic beings, just as we start as single cells. This is the case. Embryo chicks for the first two or three days after fertilization use very little oxygen, but obtain the energy which they need for growth by fermenting sugar into lactic acid, like the bacterium which turns milk sour. So do various embryo mammals, and in all probability you and I lived mainly by fermentation during the first week of our pre-natal life. The cancer cell behaves in the same way. Warburg has shown that with its embryonic habit of unrestricted growth there goes an embryonic habit of fermentation.

The first living or half-living things were probably large molecules synthesized under the influence of the sun's radiation, and only capable of reproduction in the particularly favourable medium in which they originated. Each presumably required a variety of highly specialized molecules before it could reproduce itself, and it depended on chance for a supply of them. This is the case today with most viruses, including the bacteriophage, which can grow only in presence of the complicated assortment of molecules found in a living cell.

The growth and reproduction of large molecules are not, it may be remarked, quite hypothetical processes. They occur, it would seem, in certain polymerizations which are familiar to organic chemists. In my opinion the genes in the nuclei of cells still double themselves in this way. The most familiar analogy to the process is crystallization. A crystal grows if placed in a supersaturated solution, but the precise arrangement of the molecules out of several possible arrangements depends on the arrangement found in the original crystal with which the solution is 'seeded'. The metaphor of seeding, used by chemists, points to an analogy with reproduction.

The unicellular organisms, including bacteria, which were the simplest living things known a generation ago, are far more complicated. They are organisms—that is to say, systems whose parts co-operate. Each part is specialized to a particular chemical function, and prepares chemical molecules suitable for the growth of the other parts. In consequence, the cell as a whole can usually subsist on a few types of molecule, which are transformed within it into the more complex substances needed for the growth of the parts.

The cell consists of numerous half-living chemical molecules suspended in water and enclosed in an oily film. When the whole sea was a vast chemical laboratory the conditions for the formation of such films must have been relatively favourable; but for all that life may have remained in the virus stage for many millions of years before a suitable assemblage of elementary units was brought together in the first cell. There must have been many failures, but the first successful cell had plenty of food, and an immense advantage over its competitors.

It is probable that all organisms now alive are descended from one ancestor, for the following reason. Most of our structural molecules are asymmetrical, as shown by the fact that they rotate the plane of polarized light, and often form asymmetrical crystals. But of the two possible types of any such molecule, related to one another like a right and left boot, only one is found throughout living nature. The apparent exceptions to this rule are all small molecules which are not used in the building of the large structures which display the phenomena of life. There is nothing, so far as we can see, in the nature of things to prevent the existence of looking-glass organisms built from molecules which are, so to say, the mirror images of those in our own bodies. Many of the requisite molecules have already been made in the laboratory. If life had originated independently on several occasions, such organisms would probably exist. As they do not, this event probably occurred only once, or, more probably, the descendants of the first living organism rapidly evolved far enough to overwhelm any later competitors when these arrived on the scene.

As the primitive organisms used up the foodstuffs available in the sea some of them began to perform in their own bodies the synthesis formerly performed at haphazard by the sunlight, thus ensuring a liberal supply of food. The first plants thus came into existence, living near the surface of the ocean, and making food with the aid of sunlight as do their descendants today. It is thought by many biologists that we animals are descended from them. Among the molecules in our own bodies are a number whose structure resembles that of chlorophyll, the green pigment with which the plants have harnessed the sunlight to their needs. We use them for other purposes than the plants—for example, for carrying oxygen—and we do not, of course, know whether they are, so to speak, descendants of chlorophyll or merely cousins. But since the oxygen liberated by the first plants must have killed off most of the other organisms, the former view is the more plausible.

A number of organisms exist today which cannot live in presence of oxygen. Such are the bacteria causing tetanus and gas gangrene. They may of course be descendants of air-breathers which have lost the capacity for dealing with oxygen. But I like to toy with the idea that they are the vestiges of an older order of living beings, and to think, as I examine black mud or a septic wound,

Hic genus antiquum terrae, Titania proles,
Fulmine deiecti fundo volvuntur in imo.

The above conclusions are speculative. They will remain so until living creatures have been synthesized in the biochemical laboratory. We are a long way from that goal. It was only this year[2] that Pictet for the first time made cane-sugar artificially. It is doubtful whether any enzyme has been obtained quite pure. Nevertheless I hope to live to see one made artificially. I do not think I shall behold the synthesis of anything so nearly alive as a bacteriophage or a virus, and I do not suppose that a self-contained organism will be made for centuries. Until that is done the origin of life will remain a subject for speculation. But such speculation is not idle, because it is susceptible of experimental proof or disproof.

Some people will consider it a sufficient refutation of the above theories to say that they are materialistic, and that materialism can be refuted on philosophical grounds. They are no doubt compatible with materialism, but also with other philosophical tenets. The facts are, after all, fairly plain. Just as we know of sight only in connection with a particular kind of material system called the eye, so we know only of life in connection with certain arrangements of matter, of which the biochemist can give a good, but far from complete, account. The question at issue is: 'How did the first such system on this planet originate?' This is a historical problem to which I have given a very tentative answer on the not unreasonable hypothesis that a thousand million years ago matter obeyed the same laws that it does today.

This answer is compatible, for example, with the view that pre-existent mind or spirit can associate itself with certain kinds of matter. If so, we are left with the mystery as

[2] 1928.

to why mind has so marked a preference for a particular type of colloidal organic substances. Personally I regard all attempts to describe the relation of mind to matter as rather clumsy metaphors. The biochemist knows no more, and no less, about this question than anyone else. His ignorance disqualifies him no more than the historian or the geologist from attempting to solve a historical problem.

THE BIOLOGY OF INEQUALITY

I PROPOSE to examine certain suggested applications of biology to political science. In particular I wish to examine certain statements regarding human equality and inequality, some of which have been used to justify not only ordinary policy but even wars and revolutions.

We will first consider the doctrine of the equality of man. I will quote from a great revolutionary document of the eighteenth century, the American Declaration of Independence, which was published in 1776 and is mainly due to Jefferson. 'We hold these truths to be self-evident, that all men are created equal, that they are endowed by their Creator with certain unalienable Rights, that among these are Life, Liberty, and the pursuit of Happiness.' This or a very similar doctrine of equality was important for the French Revolution. What did it mean in practice? The thirteenth and fifteenth amendments to the United States' Constitution were needed to abolish negro slavery and racial discrimination in the matter of the franchise. For whites it meant a very considerable measure of equality before the law, and it has, I think, meant a somewhat greater equality of opportunity than exists in England; but it did not give rise to any systematic attempt to bring about economic equality.

Modern revolutionary theory is much more modest in its statements regarding equality, though its practice goes somewhat further in that direction. 'The real content of the proletarian demand for equality is the demand for the abolition of classes. Any demand for equality which goes beyond that, of necessity passes into absurdity.' So wrote Engels, and the passage was considerably amplified by

Lenin. Modern revolutionary theory looks forward to two types of society; socialist society in which each citizen works according to his ability and receives in accordance with the amount of work done, and communist society in which each works according to his ability and receives according to his needs. There is a certain approximation towards socialist society in the Soviet Union, but communist society remains an ideal. Neither of these theories is equalitarian. Stalin in a report to the seventeenth Congress of the CPSU said: 'Marxism starts out with the assumption that people's tastes and requirements are not, and cannot be, equal in quality or in quantity, either in the period of Socialism or the period of Communism.' Further, so far as I know, official communist theory includes no clear statement of the origins of inequality other than economic.

Now although Jefferson regarded the truth of human equality to be self-evident there is remarkably little positive evidence for the Jeffersonian theory, and its interest is, I think, mainly historical. We shall have to consider later how much of it can be salvaged. Conservative and reactionary politicians and biologists today lay considerable stress on human inequality.

Let us now consider a series of doctrines that are based on the theory of inequality. We will take first the theory that 'the unfit should be sterilized'. I may add at once that the operation of sterilization is not castration. It means an operative interference which prevents the conception or begetting of children. It is a slight operation in men, more serious in women. There have been many statements of this doctrine. For example, Mr Justice Holmes of the United States Supreme Court, in a judgement on appeal, said: 'It is better for all the world if society can prevent those who are manifestly unfit from continuing their kind.' We must ask, however, 'Who are the unfit?' and 'Do they all continue their kind?' We must also ask who is to decide these questions, both the question of the unfitness and that of whether it is handed on; and we must ask a final question —whether sterilization is the only practicable way of preventing the individual from continuing his kind, if we find that this is desirable. There have been, of course, many attempts to answer this question, and to put sterilization on a legal basis.

I prefer not to quote the German law on the subject because it is inevitable that to do so would give rise to a certain amount of prejudice either for or against this law. I will quote the American model Sterilization Law, drafted by H. H. Laughlin in a *Report of the Psychopathic Laboratory of the Municipal Court of Chicago* (1922).[1] Here are some sections of this law; [Section 2, subsection (*a*)] 'A socially inadequate person is one who by his or her own effort, regardless of etiology or prognosis, fails chronically in comparison with normal persons, to maintain himself or herself as a useful member of the organized social life of the state; provided that the term socially inadequate shall not be applied to any person whose individual or social ineffectiveness is due to the normally expected exigencies of youth, old age, curable injuries, or temporary physical or mental illness, in case such ineffectiveness is adequately taken care of by the particular family in which it occurs.'

'(*b*) The socially inadequate classes, regardless of etiology or prognosis, are the following: (1) Feeble-minded; (2) Insane (including the psychopathic); (3) Criminalistic (including the delinquent and wayward); (4) Epileptic; (5) Inebriate (including drug habitués); (6) Diseased (including the tuberculous, the syphilitic, the leprous, and others with chronic, infectious and legally segregable diseases); (7) Blind[2] (including those with seriously impaired vision); (8) Deaf[3] (including those with seriously impaired hearing); (9)

[1] pp. 446, 447 of *Eugenical Sterilization in the United States.*

[2] For example Milton. [3] For example Beethoven.

Deformed (including the crippled); and (10) Dependent (including orphans, ne'er-do-wells, the homeless,[4] tramps,[4] and paupers[4]).'

'(*f*) A potential parent of socially inadequate offspring is a person who, regardless of his or her own physical, physiological or psychological personality, and of the nature of the germ-plasm of such person's co-parent, is a potential parent at least one-fourth of whose possible offspring, because of the certain inheritance from the said parent of one or more inferior or degenerate physical, physiological or psychological qualities would, on the average, according to the demonstrated laws of heredity, most probably function as socially inadequate persons; or at least one-half of whose possible offspring would receive from the said parent, and would carry in the germ-plasm but would not necessarily show in the personality, the genes or genes-complex for one or more inferior or degenerate physical, physiological or psychological qualities, the appearance of which quality or qualities in the personality would cause the possessor thereof to function as a socially inadequate person under the normal environment of the state.'

Now you see that goes rather far! Section 15 of the same draft law empowers the State Eugenicist to cause the potential parents of socially inadequate offspring to be sterilized in a 'skilful, safe and humane manner, and with due regard to the possible therapeutical benefits of such treatment of operation.' Such a course may be desirable. That is a matter which we shall have to discuss. I do put it forward, however, that such legislation is considerably more revolutionary than, for example, the more moderate forms of socialism, and would involve considerably more interference with individual liberty. It may be necessary in the interests of the race. That is a matter we shall have to examine later on.

[4] For example Jesus.

The third statement which we shall have to consider is that certain classes are congenitally superior to others, and that it is desirable that the superior classes should reproduce more rapidly. As an example of that type of thinking I will quote the *Report of the Committee of the Eugenics Society* (1910)[5], which commented on the *Reports of the Royal Commission on the Poor Laws* published in 1909. 'That element in pauperism which represents and transmits original defect, almost completely neglected in the investigation and wholly neglected in the recommendations of the Commission, is the one we wish to take into consideration. The determination of this element is not a matter of opinion but of the application of methods of careful investigation. It is impossible to disregard the fact that the typical dependant in the minds of the Commissioners is not the typical dependant who habitually receives relief. Yet it is precisely the latter who is primarily the subject of Poor Law relief, and who affords the chief burden on the public purse. He is not the man who responds to a call on manly independence or stands ready to take a place made available through the Labour Exchange. He was born without manly independence[6] and is unable to do a normal day's work however frequently it is offered to him.'

'In a general sense, and of course with many exceptions, the unemployed represent relatively weaker stocks. With a diminution of work elimination falls on the less qualified. This is qualified by the factor of age; elimination at forty years of age is possibly associated with elements of original weakness. If a man can do only half the work required in these days of standardized wages, it is rather futile to attempt to introduce him to the industrial system.'

[5] Quoted by E. J. Lidbetter, *Heredity and the Social Problem Group* (1933), vol. 1, p. 12.

[6] In my own experience the majority of new-born infants are devoid of this quality.

'It is to be noted finally that degenerate tendencies do not manifest in transmission a single set of characteristics but take on a great multiplicity of forms. A single family stock produces paupers, feeble-minded, alcoholics and a certain type of criminals. If an investigation could be carried out on a sufficiently large scale we believe that the greater proportion of undesirables would be found connected by a network of relationship; a few thousand family stocks probably provide this burden which the community has to bear.'

It is only fair to remember that that was written in 1910, and that since then the problem of unemployment has entirely changed in character. Those people who were regarded as unemployable have been called the 'social problem group' by the investigators of the Eugenics Society. Modern sociologists, however, rarely ascribe the unemployment of 1½ million people to congenital abnormalities. We shall later discuss the theory that many of the poor are poor because of hereditary defects.

In the same way many biologists believe in the innate superiority of certain classes and in the extreme importance of the ruling classes. Professor Fisher, for example, in his book, *The Genetical Theory of Natural Selection*,[7] writes: 'The fact of the decline of past civilizations is the most patent in history, and since brilliant periods have frequently been inaugurated, in the great centres of civilization, by the invasion of alien rulers, it is recognized that the immediate cause of decay must be the degeneration or depletion[8] of the ruling classes.' We note that Fisher takes for granted the exact opposite of the proposition concerning equality that was obvious to Jefferson. The truth may lie between the two ideas.

[7] p. 237.

[8] Thus if English culture decayed during the sixteenth century and particularly under Elizabeth, this may be attributed to the depletion of the feudal nobility in the Wars of the Roses,. If not, not.

We have next to consider the fourth doctrine 'that certain races are congenitally superior to others'. The earliest statement of that doctrine known to me is found in the Book of Genesis, where the curse on the children of Ham is related. It is worthy of note that if this attribution of priority is accurate, the doctrine of racial superiority is originally a Jewish doctrine, although it is now being used against the Jews in Central Europe. There have been some very surprising statements of this doctrine by recent German authors since 1933. I will only quote one to give an example of the remarkable theories current in Germany. Dr Johann von Leers writes: 'After a period of decadence and race obliteration, we are now coming to a period of purification and development which will decide a new epoch in the history of the world. If we look back on the thousands of years behind us we find that we have arrived again near the great and eternal order experienced by our forefathers. World history does not go forward in a straight line but moves in curves. From the summit of the original Nordic culture of the Stone Age, we have passed through the deep valley of centuries of decadence, only to rise once more to a new height. This height will not be less than the one once abandoned, but greater, and that not only in the external goods of life.'

It is interesting to think that the Nordic race, if properly purified, may rise even higher than the culture of the Stone Age. When one reads such statements as that, one is tempted to ask whether they are made in order to obtain or retain posts, or whether, possibly, they may not be a rather subtle form of propaganda intended to make the existing racial doctrines in Germany appear ridiculous. I therefore propose to quote from German authors who wrote before the advent to power of the National Socialist Party, and who therefore expressed themselves much more moderately. For example, E. Fischer, writing in 1923, gives the following account of the Nordic race: 'The mental endowments of the Nordic race are great energy and industry, vigorous imagination, and high intelligence. Conjoint with these are foresight, organizing ability and artistic capacity (this being least marked in respect of music); and also the unfavourable qualities of strong individualism, a lack of community sense and of willingness to obey orders, a certain one-sidedness and an undue inclination towards imagination and flights of fancy, a dislike for steady and quiet work; while as additional qualities may be mentioned a considerable expansive force, a power of devotion to a plan or idea, an adequate capacity for instilling an idea into others and a small inclination for adopting the ideas of others—in a word significant powers of suggestion and comparatively little suggsestibility. It is obvious that when circumstances are favourable persons richly endowed with these gifts are likely to become leaders, inventors, artists, judges and organizers.' These admirable qualities are regarded as inherent in the Nordic race, and it is thought that the regeneration of Germany can only come from the spread of the Nordic elements in it. For example, Günther in 1929 wrote: 'The rise of the peoples of Germanic speech is given by the increase of the healthy hereditary units, and an increase of Nordic blood.'

Later, however, there has been a tendency to speak not so much of the Nordic race but of the German race. We shall have to ask whether a Nordic or German race exists, and if so whether these doctrines as to natural endowments can be justified.

The fifth doctrine is that 'crossing between different races is harmful'. Lenz wrote: 'In my opinion there can be no doubt that the mingling of races widely distinct from each other may lead to the production of types which are disharmonious in respect alike of body and mind.' More recent German statements of the terrible effects of race mixture will be familiar to you. As the most striking of them, those of Herr Streicher in *Der Stürmer*, are highly obscene, I do not propose to prejudice the minds of readers by quoting them.

Now before we can examine these and related theories in detail we must consider the general biology of inequality. That is dealt with by the science of genetics. Genetics is primarily concerned with innate inequality, but has to consider all kinds of inequality, or as we call it in biological terminology 'variation'. The success of genetics, which deals among other things with heredity, has been due to the fact that during the last forty years it has been concerned with differences rather than with resemblances. It is possible to give a reasonable answer to the question, 'Why is this mouse black while this other mouse is white?' but at present it is not possible to give at all an adequate answer to the question, 'Why are they both mice? Why does a pair of mice produce another mouse and not a rat or a motor bicycle?' By concentrating on these relatively small differences genetics has advanced a considerable distance, and it is therefore peculiarly adapted to deal with this problem of human inequality. I am certainly not going to answer such a question as 'What is Man?' I may be able to help you to get a clearer view as to the nature of the differences between individual men and their causation. If so I may be able to answer at least provisionally some of the questions which I have posed already.

Let us suppose that we have before us two dogs, both with legs somewhat bent. It may be that one of these dogs has bent legs because as a puppy it received a diet containing inadequate quantities of antirachitic vitamin, while the other dog has bent legs because its father was a dachshund. In the first case we say that the difference between this dog and a straight-legged dog is due to nurture. In the second case we say that it is due to nature. It is possible by experimental methods with animals and plants to separate these two causes of variation, but we must recognize that in the majority of cases both of the causes are operative. If we are dealing with the sizes or weights of a number of animals we shall certainly find that the differences are due both to differences of nature and to differences of nurture. In man, where experiment is impossible, we shall find it very much harder to determine the origin of the differences that undoubtedly exist. Let us see what happens when we tackle the problem experimentally.

If in any branch of science we find that one quantity or quality varies as a function of several others we shall design our experiment so as to keep all but one of the independent variables approximately constant. If, for example, we wish to ascertain the laws governing the volume of a gas we shall first keep the temperature constant while varying the pressure and so discover Boyle's law. If we keep the pressure constant, while varying the temperature, we shall discover Charles's law. If we begin by measuring the volume of gas at an arbitrary series of temperatures and pressures we shall find our work very much more difficult.

Now we do in practice try to eliminate our variables —our differences of nature and of nurture. I have not defined them because we shall be able to understand their essence very much more clearly when we deal with the practical methods that render them uniform. The most obvious thing to do is to make the nurture of our various organisms as like as possible. If we are growing a number of plants we shall see that they all have the same soil, the same amount of water and light, and that the density of plants in one part of our field is the same as in the other. With animals we shall take similar precautions. For example, we shall be particularly careful that if there is any infection all members of the population shall be equally exposed, and if we were dealing with man in an ideal experiment we should have to try to render the education and social environment of our different individuals as uniform as possible. We should then see that any differences that remained in that uniform environment were probably due to nature and not nurture.

I am perfectly aware that a uniform environment is an impossible ideal, nevertheless it is easy to think of characters which are very slightly affected by the environment, for instance eye-colour in man. It is easy to think of other characters, which, although they vary considerably with the environment, may quite readily be stabilized by making the environment similar—for example the skin colour which varies a good deal according to the amount of sunlight to which a man is exposed. We may take it that during a winter in England there will not be very much sunburn.

We see then that it is possible to a large extent to eliminate one of our variables, nurture, at least when doing experimental work with plants and animals. How shall we perform the converse operation? How shall we get a population of animals or plants uniform as regards their nature, their innate qualities? There are three ways of doing so. First of all we may grow what is called a clone, that is to say a population of individuals which are derived from the same individual by vegetative reproduction. For example, if you buy a named variety of a potato, tulip, rose, or apple, you will find, if your seedsman is an honest man, that the plants which you buy have all been derived from the same seedling by vegetative reproduction. The original potato was divided; it sent out roots on which new tubers were grown, and these were used as so-called seed potatoes for the next generation, but there was no sexual reproduction. If we take a well-known type such as Arran Victory we shall find that it produces a large variety of different potatoes if its seeds are sown. In the case of a named rose we propagate it by grafting but there is no sexual reproduction. Within a clone we find considerable uniformity, and in so far as there are differences they are not in general handed down. Selection within a clone is ineffective. If you once have your named variety of rose, except for a very occasional bud sport, you will not improve it by selecting the best individuals from it. Such differences as exist appear to be temporary effects of environment which are not transmitted.

One may ask, 'What has that to do with man? Man does not reproduce vegetatively.' In human beings there are two types of reproduction, the ordinary sexual reproduction, and much more rarely asexual reproduction. The embryo in its early stages may divide to give a pair of monozygotic twins, who resemble one another to a very remarkable degree, and who are believed, on genetical grounds that seem to me entirely sound, to have the same nature. They have, of course, very much the same nurture up to the time of birth and often afterwards.

A second type of genetically uniform population is that called a 'pure line'. You may get that by self-fertilizing a plant for ten generations, or in the case of animals by brother and sister mating, for a larger number—thirty or more—of generations. Such a pure line is generally very uniform. There may, however, be differences within it.

The remarkable point is that these differences are not inherited. It is easy to breed pure lines of the fly called *Drosophila funebris*, since it accomplishes a generation in about twenty days, and one can breed 400 in a single bottle. In the normal type of fly the veins extend to the margin of the wing. In some abnormal individuals one is broken. It is possible by suitable crossings to produce populations in which a given proportion of the individuals have that vein interrupted. It may be very few, it may be 100 per cent; it may be intermediate. It is possible by continued brother-sister mating to obtain a pure line in which the proportion of abnormals is the same in all families.

Table 1 gives the actual figures obtained by Timoféeff-Ressowski in such a line.

TABLE 1

Offspring of Abnormal Flies		Offspring of Normal Flies	
Abnormal	Normal	Abnormal	Normal
199	24	311	36
288	32	201	25
192	22	219	25
679	78 = 10.3 ± 1.1%	731	86 = 10.5 ± 1.0%

It will be seen that in this line 10 per cent of the individuals had normal wings, and 90 per cent abnormal. These proportions were the same whether he bred from two parents with normal wings or two with broken veins, provided both were members of the line. Selection for normality or abnormality is completely ineffective. That is to say, although there are differences with regard to the wing vein those differences are not inherited. What is inherited is a constitution such that in a particular environment 10 per cent have a normal wing and the remainder have a broken vein. I am aware that that is a somewhat difficult conception to grasp. It is fundamental in modern genetics. We cannot always speak of the inheritance of a character; in many cases we can speak of the inheritance of a constitution which in a particular environment will give such and such a range of characters.

Now within a pure line all differences, as far as we can see, are due to nurture, none to nature. If we alter the conditions so that a larger proportion of our flies have a particular character, that character is not handed on to the offspring when the original environment is restored. As a result of such experiments very few geneticists nowadays believe in Lamarck's doctrine that 'acquired characters are inherited'.

The third type of genetically uniform population which we can get is the first cross between two pure lines. A second generation is generally very variable, but the first cross is often uniform and considerably more vigorous than either of the lines.

A study of pure lines teaches us that there is a certain residual variation which we cannot eliminate, even if we eliminate all differences of heredity. It is possible that if we could get an absolutely uniform environment we could eliminate these differences also. In an environment as uniform as we can get we shall still find them.

In an ordinary population, for example in any human population, there are no pure lines—a point of very considerable importance. A pure line, however, is not merely a laboratory curiosity. The named varieties of many seed-plants, for example wheat or sweet peas, approximate very closely to pure lines. Although, therefore, the pure line has no immediate applicability to human problems it can give us a great deal of information. For example, we are apt to think of congenital qualities in a baby, qualities with which it is born, as being probably due to nature, likely to depend on the make-up of its parents, and likely to be transmitted to the offspring. Let us see how far that is true by considering a particular character as manifested in four pure lines of guinea-pig.

Guinea-pigs quite frequently have extra toes. By suitable selection combined with inbreeding it is possible to produce a pure line in which the frequency of extra toes may vary from 12 per cent to 56 per cent in particular cases. You will

TABLE 2

PERCENTAGES OF POLYDACTYLOUS GUINEA-PIGS

Age of Mother	Line A	Line B	Line C	Line D
3–6 months	29.3	34.6	68.1	81.0
6–9 months	7.4	28.2	54.4	69.5
9–15 months	9.6	21.9	28.9	50.0
15– months	6.1	12.1	22.0	30.2

Effects of heredity and environment on the frequency of polydactyly (extra toes) in 1,986 guinea-pigs. After Wright.

not get any more extra toes from the extra-toed members of one of these lines than from the normal ones. The percentage of extra toes represents the reaction of that line to its particular environment. We next ask what is the most important element in the environment determining extra toes. In the particular set of environments met with in Wright's work by far the most important element was the age of the mother. You will see that in line A the young mothers produced 20 per cent of offspring with extra toes and the old mothers only 6 per cent. In line D the young mothers under six months produced 81 per cent and the older ones over fifteen months only 30 per cent. This at once shows that a character can be determined to a considerable extent both by heredity and by environment. The differences between these four lines are, of course, hereditary. The differences between the different rows in the diagram are environmental. If one is a rabid environmentalist one will read that table from top to bottom, if a rabid eugenist, from side to side. If one is a biologist one will read it both ways.

One type of human mental defect is determined in this way, namely, 'Mongolian imbecility', a condition in which, as Penrose has conclusively shown, the age of the mother is an important determining factor. The average age of the mothers of these imbecile children is about thirty-nine years, whilst that of mothers of normal children is under thirty. Besides this environmental factor there is a genetical factor, as is shown by the fact that two mothers of such imbeciles are often related. This means that an embryo of a certain constitution will develop into a Mongolian imbecile in a particular type of pre-natal environment provided by an elderly mother, or that a mother of a certain constitution provides a special type of pre-natal environment when she ages. There is some evidence that a few other kinds of mental defect are determined in a similar manner. On the other hand mental defect due to injury at birth seems to be commoner in first-born children, who are generally brought into the world with more difficulty than later children.

It is frequently asked, 'What is the relative importance of nature and nurture?' That is a question to which no general answer can be given. It is obvious that if in the population of guinea-pigs no female were allowed to breed until she was six months old the differences due to nurture would be considerably reduced. If the population had contained only three lines instead of four the differences due to nurture would have been diminished. It is possible, by suitable choosing of your character, your population, and your environment, to produce a population in which a given character is determined entirely by differences of nature or entirely by differences of nurture, and therefore the question has to be answered separately for any given population and any given character. For example, we may take such a character as illiteracy and we may compare the amount of illiteracy in adults under forty in England and India. In England we should find that the people who could not read were almost all either blind or mental defectives. We should find reason to believe that a considerable amount of the blindness and mental defect in England was due to differences of nature. On the other hand, if we went to India we should find that the majority of the illiterates were illiterate because they had had no opportunities for learning to read, and therefore differences in that respect were mainly matters of nurture. One could give, of course, many more examples of the same kind. The important point is to realize that the question of the relative importance of nature and nurture has no general answer, but that it has a very large number of particular answers.

It is fortunate for our purpose that although pure lines do not exist in man there are nevertheless human groups which breed true, or very nearly true, for certain physical characteristics. For example, we can be reasonably sure that the skin colour of the children of two English people will vary between fairly narrow limits, while the children of negroes will vary between other limits, but there will be no overlapping. We shall have to consider whether there is evidence for the existence of psychological characters that are equally closely determined.

Before we do that we must consider the interaction of nature and nurture. Let us suppose we have two different stocks which are pure lines or at least do not have very great innate variation as regards the particular character which we are studying. It may be a physical character such as weight, a physiological character such as milk yield, or some of the numerous forms of human achievement. But we will suppose that we can order our populations as regards their achievement. We can probably say that this group is on the average significantly heavier than that group. By 'significantly' I mean that the difference is such that it cannot well be due to sampling error. We may say with regard to a particular intelligence test, 'This group does significantly better than that one', and that is a statement we can make quite regardless of the philosophical question whether intelligence can be measured.

Now suppose that we have two races *A* and *B* in two environments *X* and *Y*. And suppose that we have samples of each race in each environment sufficiently large to enable us to order them without doubt as to their achievement in some respect, say longevity, milk yield, or intelligence. If their achievements overlap, we can still order them with certainty by taking large enough samples.

TABLE 3

		X	*Y*			*X*	*Y*
1.	*A*	1	2	*or*	*A*	1	3
	B	3	4		*B*	2	4
		X	*Y*				
2.	*A*	1	4				
	B	2	3				
		X	*Y*				
3.	*A*	1	2				
	B	4	3				
		X	*Y*			*X*	*Y*
4.	*A*	1	3	*or*	*A*	1	4
	B	4	2		*B*	3	2

There are exactly four possibilities, shown in Table 3. The enumeration is so simple that no one has ever troubled to make it. Nevertheless, I believe it is worth making. In the first type of interaction race *A* is superior to race *B* in each environment, and environment *X* is more favourable than environment *Y* to each race. The numbers 1, 2, 3, 4 denote the order of achievement of the four populations. This is a common type of interaction. It would be exemplified if we took two races of dog, say mastiffs and dachshunds, as races *A* and *B*, and a good diet and a starvation diet as environments *X* and *Y*. It is clear that on the better diet each race of dog would be heavier than on the poor diet. But on each diet the mastiffs would be heavier than the dachshunds. If nature and nurture always interacted in this way we could say with scientific accuracy, 'This is a heavier race of dog.' 'This is a more musical race of men.' 'This is a more fertile breed of poultry.' 'This is a healthier environment than that.' But unfortunately, things are not always so simple in reality.

Now consider the second kind of interaction. Let *A* be Jersey cattle and *B* Highland cattle. Let *X* be a Wiltshire dairy meadow, and *Y* a Scottish moor. On the English pasture the Jersey cow will give a great deal more milk than the Highland cow. But on the Scottish moor the order will probably be reversed. The Highland cow will give less milk than in England. But the Jersey cow will probably give less still. In fact, it is very likely that she will give none at all. She will lie down and die. You cannot say that the Jersey is a better milk-yielder. You can only say that she is a better milk-yielder in a favourable environment, and that the response of her milk yield to changes of environment is larger than that of the Highland cow. Our specialized domestic animals and plants generally behave in this way. It is likely that certain human types react in a similar manner.

It is, of course, possible that the interaction between nature and nurture is of a simpler type in the determination of human intelligence than in that of the milk yield of cattle or the seed yield of wheat plants. But even a thoroughgoing materialist might well doubt this. Unless it is true we cannot in general say that *A* has a greater innate ability than *B*. *A* might do better in environment *X*, and *B* in environment *Y*. Had I been born in a Glasgow slum I should very probably have become a chronic drunkard, and if so I might by now be a good deal less intelligent than many men of a stabler temperament but less possibilities of intellectual achievement in a favourable environment. If this is so it is clearly misleading to speak of the inheritance of intellectual ability. This does not mean that we must give up the analysis of its determination in despair. It means that the task will be harder than many people believe.

The third type of interaction may be illustrated by normal (*A*) and mentally defective (*B*) human children. The normal children will do best in an ordinary school (*X*), but even in a special school (*Y*) for mental defectives they will do better than defective children. On the other hand, the defective children will do better in the special school. In each environment the normal children will be superior. But the environment which is better for the normal child will be worse for the defective child, and conversely. We could equally well illustrate our case if *A* were normal bean plants and *B* a race of beans which turn white in strong light, while *X* was full sunlight, and *Y* partial darkness.

As an example of the fourth type let *A* be Englishmen and *B* West African negroes. Let *X* be an English town and *Y* the Gold Coast colony. Let the four populations be placed in order of their average lengths of life. We should probably find that the order was: English in England, negroes in Africa, negroes in England, English in Africa. We should certainly find that each race lives longer in its native environment than when transplanted. We could not say that as regards health as measured by longevity either race or either environment was superior to the other. The Englishman in West Africa is killed off by yellow fever, the negro in England by tuberculosis, each having a considerable immunity to the disease prevalent in his native land.

If we merely have two races and two environments Table 3 exhausts the possibilities unless two or more of the four achievements are equal. With a number of races and environments things are of course more complicated. But after studying Table 3 we shall be a little suspicious of such phrases as 'a good heredity', 'a good environment', or 'a superior race'. Unfortunately, almost all current theory is based on the view that the first type of interaction is universal, and this applies equally to the supporters and to the critics of eugenics.

In a mixed population things are not so simple. We may find populations in which most of the differences are due to heredity in the strict sense of a resemblance between parent and offspring. If you ask why a given dog is a greyhound, it will be correct to answer 'because both his parents and his ancestors for some way back were greyhounds'. If you ask why a given cat is tabby it will not usually be accurate to say 'because both his parents were tabbies'. In that respect the cat population presents a closer analogy to the human population than do the dogs. Let us try and see what we have to deal with besides heredity as a cause of innate differences, differences of nature. Suppose we cross a pure-bred black rabbit with a pure-bred blue rabbit, the hybrids will be black. But if we cross the hybrids together we shall get some blacks and some blues. The differences between the blacks and the blues are differences of nature, because, unlike differences due to nurture, they are handed down to the offspring. The process by which the black rabbits give rise to blacks and blues is called segregation. We shall examine it in greater detail later on. It must be carefully distinguished from the effects of environmental differences which are not transmitted to the offspring. The kinds of differences which we may get within a human population are summarized in Table 4.

First of all there are differences that are due to differences of nurture—the difference between a sunburned child and a child that is not sunburned; and between a normal child and a rickety child. Secondly, there are differences of nature, which fall into three categories; differences of ancestry, for example, the difference between a negro child and a white child, which is due to heredity; differences between brown-eyed and blue-eyed sibs, and in general all heritable

TABLE 4

CAUSES OF HUMAN INEQUALITY

NURTURE	Differences due to different environments.
NATURE	Heredity. Differences of ancestry. Segregation. Differences due to chance combinations of genes. Mutation. Changes in genes.
X	

differences between brothers and sisters which are due to segregation; differences due to mutation, a rare event of considerable biological importance. I leave a blank space X for differences which cannot be ascribed to any of these. If there is such a thing as freedom of the will in the more extreme sense that comes under X. I regard it as unscientific to leave out X, if only for this reason, that if there is no such thing as X, if all differences between human beings are strictly determined, then it should be possible in the course of some centuries to prove that, let us say, 99.9 per cent at least of all differences of certain kinds are determined by differences of nature or nurture. To my mind a proof that 99.9 per cent were so determined would be very much more effective than an assertion on a priori grounds that 100 per cent were so determined. If therefore we leave X in our table we can say that in certain cases, for example that of skin colour, X is fairly small, and we may hope according to our philosophical views to prove either that X is negligible or considerable as regards differences of conduct.

This may be regarded as a prolegomenon to any systematic treatment of human inequality.

BEYOND DARWIN

DARWIN taught that the direction of evolution was determined mainly by the survival of the fittest. Animals of the same species differ among themselves. One mouse can run faster than the average. Another has better hearing. Still another has sharper teeth. These differences are at least partly inherited. And as there is not room in the world for all the mice born, the fittest on the whole survive, and thus the species gradually changes.

A character which is useful in one environment may be harmful in another. Thus thick fur is useful in the Arctic and harmful in the tropics. Wings are generally useful to an insect in the middle of a continent, but dangerous on small islands in the ocean, where winged insects are blown out to sea, but wingless ones survive. This is one of the ways in which a species divides into two or more new species.

Marx and Engels accepted this theory of the struggle for life 'as the first, temporary, incomplete expression of a recently discovered fact'. They pointed out that 'Darwin discovers among plants and animals his English society' based on unrestricted competition. And, of course, since Darwin's time many theorists have tried to justify cut-throat competition and the oppression of the weak in the name of Darwinism.

But Marx and Engels did not deny the struggle for life in nature because they thought that men could and should behave better than animals. Kropotkin wrote of co-operation in nature even between different species. This occurs, but it is exceptional.

The dialectical method in science is to push a theory to its logical conclusion, and show that it negates itself. For example, we know that the so-called atoms of chemical elements are not really indivisible. But this would never have been discovered if chemists had not believed in the existence of atoms, and investigated their properties with great care. Dalton's atomic theory is still the basis of chemistry. But it is such a good theory that it disproves itself, and makes way for a nearer approach to absolute truth.

It is the same with Darwinism. Animals and plants are not quite such ruthlessly efficient strugglers as they would be if Darwinism were the whole truth. It is true that a lot of what at first seems useless beauty is part of the struggle. Thus flowers are useful to plants because they attract insects. And they are beautiful to us because we share the aesthetic preferences of insects to some extent.

However, it has recently been shown that the struggle for life defeats itself if it is pushed too far. So long as a species is mainly struggling against other species or external nature, it usually becomes fitter. But when the struggle occurs within a species this is not so. Thus if male animals fight for females, the most successful fighters will have most children. So the species may develop weapons and instincts which are only useful in fighting their own kind.

In particular, mere size is an advantage in such struggles. Animals where the male is much larger than the female, such as the domestic fowl, the sea elephant, and many species of deer, are generally polygamous. Whereas in animals with monogamous families, such as most birds, the sexes are generally of the same size. And the study of fossils shows that a steady increase in size generally ends in extinction. Large animals are usually less fit than small ones for flying, burrowing, making their way through thick vegetation, walking on boggy ground, and in many other situations.

The Soviet biologist Gause has studied the struggle for

existence among small animals and plants in aquarium tanks. If we put in two species one of which eats the other, the eaters increase until their prey diminishes in numbers, and then begin to die of starvation. The numbers fluctuate periodically, like the numbers of men employed under capitalism.

And a crisis may become so acute that first the eaten and then the eaters die out. To prevent this, it may be necessary that the prey should have some kind of shelter where the eaters cannot find them. In fact, too great efficiency may lead to extinction. It is very unusual for a herbivorous animal to eat up its food plants completely. But this sometimes happens, as when a plague of caterpillars completely strip the oak-trees in a region, and the caterpillars then die of starvation.

The Oxford biologist Elton has given many examples of this principle. For example, the Red Indians in Labrador used to hunt caribou with spears and other primitive weapons. When they got guns they killed off so many deer that they starved or were compelled to buy imported food and live in settlements, where they caught European diseases.

Elton takes the view that it does not always pay a species to be too well adapted. A variation making for too great efficiency may cause a species to destroy its food and starve itself to death. This very important principle may explain a good deal of the diversity in nature, and the fact that most species have some characters which cannot be accounted for on orthodox Darwinian lines.

Elton is not, so far as I know, a Marxist. But I am sure that Marx would have approved of his dialectical thinking, and that it is on such lines that the Darwinian theory will develop. I do not think that Darwinism will be disproved. But it will certainly be transformed.

THE STRANGE CASE OF RAHMAN BEY

IN several daily newspapers of 1 July 1938, there was a story of an Egyptian called Rahman Bey, who threw himself into a trance, and stayed for an hour at the bottom of a swimming-bath in a metal tank. 'My peculiar gift was discovered by a Yogi priest when I was a child', he is reported to have said. And the whole thing is put across to the British public as a sample of the mysterious gifts of Orientals, who, of course, are so unlike us!

Now there are two funny things about this story, before we come to the tank at all. The name Rahman Bey means Colonel Merciful, which strikes me as funny. And as the Yogi philosophy is a product of India, it is remarkable that its adherents should experiment on Egyptian children.

The *Daily Herald* gave a photograph of the tank, which measures 8 feet by 1½ feet by 2 feet, according to one of my colleagues, and rather less according to me. So it holds 20 to 25 cubic feet. An average man occupies 2½ cubic feet, so Rahman Bey had about 20 cubic feet of air. Now a man doing light work uses about 24 cubic feet of oxygen in 24 hours. If he lies still, this is reduced to about half.

So Colonel Merciful used about half a cubic foot of oxygen in an hour. As 20 cubic feet of air contain just over 4 cubic feet of oxygen, he had plenty to spare at the end of the hour, and could have gone on for another two hours. After this time he would have been very short of breath, and would have panted so much that the remaining oxygen would have been used up more quickly. And when he came out after three hours he would have had a nasty headache. For besides using oxygen, a man makes a slightly smaller amount of carbon dioxide, and after breathing air containing anything over 6 per cent of this gas for an hour, one has a short but violent headache, and I for one sometimes vomit.[1]

Before being shut up in the tank, Rahman Bey 'shook like a pneumatic drill and then flung himself violently into unconsciousness'. I should have lain down quietly. But if Rahman Bey was unconscious he saved himself an hour of boredom. However, if I can borrow a tank, I am perfectly willing to spend an hour at the bottom of a swimming-bath for a suitable fee to be paid to the International Brigade Dependants Fund.[2]

Many readers will say 'What does all this matter?' It matters quite a lot. The physiology of human breathing is involved in questions of mine and factory ventilation, and of protection against poison gas, which today concerns everybody.[3] And because we are not educated in the matter, we take the government's statements on gas defence as seriously as the journalists took Rahman Bey.

I recently had an airtight tank with a glass window made in which a child could be shut up for several hours during a gas attack. I stayed in it for an hour myself, and got rather warm, whereas Rahman Bey was doubtless cool at the bottom of his swimming-bath. But I had to try three mothers before any of them would allow their baby to stay in it for even half an hour.

There are no gas masks for babies,[4] and a tank of this kind would give protection for some hours at any rate. But because we are not taught such elementary facts about ourselves as how much oxygen a man uses per hour, a good many babies are going to die if gas is used against civilians in a future war.

[1] So do others, as the experiments on the International Brigaders in connection with the loss of the *Thetis* showed.

[2] Unfortunately, this challenge was not taken up.

[3] As the *Thetis* case showed, it is also important to submarine crews.

[4] A few are now available.

You may call me a crass materialist, but it seems to me more important that children should be taught such facts as this than that they should know how often King Henry VIII married or who won the Battle of Agincourt.

There are a great many strange stories about the wonderful powers displayed by various Asiatics and Africans. When they are investigated they generally turn out either to be untrue or to be based on elementary facts of human physiology which are known to certain groups in India, but not yet generally known in Europe.

For example, I have pushed a red-hot cigarette end against the finger of a hypnotized Englishman without causing either pain or blistering. If he had been an Indian, it would have made a story for the daily Press.

These stories are very useful to imperialists, because they help to spread the idea that the human races are very different. If people in England believe the myth that members of coloured races have powers which Europeans do not possess, they will be ready to believe another myth —namely, that they do not possess the power of looking after their own affairs.

It is time that we realized that scientific investigation has shown that people of different races are remarkably alike, and that it is only prejudice and the self-interest of exploiters which prevent them from being brothers.

BLOOD AND IRON

WHY do we need blood? An average man has about a gallon of blood, and is likely to die if he loses half of it. The blood serves many functions. It carries food materials from the intestine to the rest of the body. It carries water and waste products to the kidneys. But its most urgent function is to carry gas; oxygen from the lungs to the muscles and other working organs, and carbon dioxide from the muscles to the lungs. Ordinary fluids take up very little oxygen, so that they would be no good for the purpose. Blood contains a special substance, haemoglobin, which combines with oxygen very easily, and gives it up easily as well. In fact, a pint of blood carries nearly as much oxygen as a pint of air.

Haemoglobin is of a deep purple colour. If you want to see the colour, prick your finger or ear lobe and let it bleed into some water till you have a nice clear red fluid. Put this in a small bottle and add a crystal of sodium hydrosulphite (not hyposulphite). This will combine with the oxygen and the liquid will turn purple.

Shake it up with air or bubble air through, and it goes red again. This is what happens several times a minute in your own body. The blood in the vein from a working muscle is almost black. Now bubble a little coal gas through the blood. It goes pink and stays pink, even if you add sodium hydrosulphite. The haemoglobin is combined with carbon monoxide, and is useless as a carrier of oxygen. If you put your head in the gas oven your lips will be a nice pink colour after you are dead. But blood that stays red is of no use to you, and you will have died because it stayed red.

Haemoglobin is a protein—that is to say, consists of large molecules much like those found in meat, eggs, cheese, and other foodstuffs. But, unlike most proteins, it contains iron. Not a great deal of iron. You have less than $\frac{1}{10}$ oz. of iron in your whole body. Nevertheless, you may easily go short of iron.

You make your haemoglobin in a very curious place, your bone-marrow. The haemoglobin is carried about in red corpuscles, which are too small to see with the naked eye, but can easily be seen with a microscope, and just seen with a very powerful magnifying glass.

They last for about a month, and then wear out, and are scrapped in the liver. The iron is mostly carried back to the marrow. The coloured part of the haemoglobin is thrown out in the bile and finally got rid of in the excreta. If the bile-duct is blocked it goes into the blood, and you get jaundice, for the removal of iron and other changes have altered the colour from red to yellow.

If you have too little haemoglobin you are anaemic. When you work, your muscles cannot get enough oxygen, and you become weak and short of breath.

Anaemia is one of the commonest diseases. In the tropics it is often caused by a small worm, *Ankylostoma*, of which hundreds may find their way into the human intestine, where they suck the blood. This disease is very common in tropical countries where the eggs of these worms live in polluted mud, and bore their way into the legs of bare-footed men, women, and children. This can be completely prevented by proper sanitation and disposal of refuse.

The hookworm, as it is called in the United States, was apparently brought over by negro slaves. They lived under filthy conditions, and infected one another and their white masters. A few years ago it was found that a large proportion of the whites in the south-eastern states suffer from anaemia due to hookworms. And it is quite possible that this worm played a big part in winning the Civil War, in which the slave-owners were beaten. The slave-owners kept their slaves without proper sanitation, and the slaves took their unconscious revenge by giving their masters anaemia.

About 1900 there was an outbreak of this worm disease in the Cornish tin-mines, where the climate underground is tropical. It was soon stopped by installing proper sanitation.

In England we are just beginning to discover how common anaemia is. Forty years ago my father invented an apparatus by which the amount of haemoglobin in a drop of blood could be accurately measured. He bled a number of men, women, and children, of whom I was one of the youngest, and worked out the averages. He found that women had less haemoglobin than men, with very few exceptions.

For a long time doctors thought that this was a natural peculiarity of women, like their smaller average height. But in 1936 Drs McCance and Widdowson, of King's College Hospital, London, found that women, even in the well-to-do classes, were chronically short of iron. The majority of them made more haemoglobin if they were given more iron in their diet.

Like many other characters in which women are supposed to be inferior to men, this one turned out to be mainly due to external conditions, and not to be an inborn defect. Women lose some blood every month, and naturally need more iron in their diet than men. But they generally get less.

The iron can be made up in several ways. Many foods contain iron, but, especially in meat, much of the iron is in an indigestible form. The best-known sources are liver, cocoa, and winkles. But parsley, haricot beans, peas, and lentils are also rich in iron. Brown bread and eggs are also good sources. In spite of Popeye the Sailor, spinach does not rank very high; and milk, which is otherwise an excellent food, is very poor in iron, though better than beer.

Many medicines sold as cures for anaemia contain digestible iron salts and are therefore valuable. But usually a 2s. bottle of medicine contains about a farthing's worth of iron salts, so you will do better to spend the money on liver or cocoa.

Other things than iron are needed to make new blood. A fairly common disease in India, tropical aplastic anaemia, is due to the lack of a substance whose exact nature is at present unknown, which is missing in the diet of many poorly paid Indian workers.

Many other substances besides iron are deficient in a lot of British diets, particularly among the workers. But iron is the very simplest of dietary needs, after water, and a study of our needs of iron is a good introduction to the general theory of diet.

GERM-KILLERS

ONE thing which I particularly liked about the Soviet Union when I visited it was the almost complete absence of propaganda as compared with England. I don't mean political propaganda, but biological propaganda, propaganda which disgusts me as a biologist. In the advertisement columns of the newspapers and on the hoardings we are told the most fantastic biological tales. For example, that it is dangerous to have acid in your stomach, that pains in the back are generally due to kidney disease, or that your skin is healthier if you smell like a tar-barrel instead of a human being.

In truth, you need strong acid in your stomach for normal digestion; and pains in the back are generally due to inflammation of the back muscles, displacement of the joints in the back, or disease of some other internal organ than the kidney. And while an occasional bath is very desirable in order not to get lousy, too much washing with soap appears to remove a protective film of grease which is secreted by special glands in the skin.

This propaganda is even more obviously put over for economic reasons than the very similar propaganda about 'our' Empire, and 'our' flag, which, by the way, no longer serves to protect British seamen, as it once did. In this country one can still reply to some of the political propaganda. But one cannot reply directly to the biological propaganda. I was brought up on the comic theory that our Press is free.

About fifteen years ago I was writing a book in which I mentioned the analysis of Beecham's Pills which had been made on behalf of the British Medical Association, and published in their book, *Secret Remedies*. The publishers at once cut the passage out. They did not want to risk an action for libel, though I never suggested that these pills contained any harmful substance, but only gave the estimated price of their ingredients. So in this article I shall only be able to write a very tiny fraction of the truth.

The medicines which you buy at the chemist's are of two kinds. They may be of secret composition, in which case you will generally find a stamp for taxation purposes on the packet. This sort of medicine usually contains a well-known drug puffed up as a wonderful secret discovery, but sometimes consists of sugar and water with a pink dye, or diluted vinegar. A few are definitely harmful. Some of the slimming cures act by producing such violent indigestion as to force you to starve yourself.

The other medicines give their composition, and many of them contain well-tried remedies, which work in a great many cases. The formulae are often less mysterious than they look. Thus 'Koray', which is well known to readers of the *Daily Worker*, consists mainly of acetyl-salicylic acid, put up with starch (amylum), cane-sugar (sucrose), and a small amount (*q.s.* is an abbreviation for the Latin *quantum sufficit*, or 'as much as is needed') of tetrabromfluorescein, a red colouring matter. You actually get about half as much again of the acetyl-salicylic acid, which is the active constituent, per tablet as in some other preparations of the same kind.

The mystery is partly the fault of the medical profession. They have a series of most impressive-looking symbols for weights and measures, and use Latin words for the most ordinary substances. For example, glycirrhiza means liquorice, and if you ask your doctor to prescribe for a child's constipation, he is likely to prescribe 'Extractum glycirrhizæ liquidum', which is much more impressive than 'liquorice and water', and also costs much more.

In spite of this, about once in five times the doctor finds something seriously wrong, with which he can deal, and with which you could not deal yourself. So we cannot dispense with doctors just because they do not always tell us the whole truth.

In fact, there is a medical racket and a 'patent medicine' racket. Each fiercely denounces the other. The doctor tells you to avoid proprietary articles, and often prescribes the same substances himself. The patent medicine merchant tells you to avoid harmful drugs and trust to natural remedies, although he may be selling a natural product which is quite a powerful drug and harmful if you take too much.

This racketeering is inevitable under capitalism for two reasons. In the first place there are big profits to be made out of selling either genuine or bogus medicine. Secondly, the general public is not taught medicine or pharmacology. This need not be so. Lenin looked forward 'to the education, training, and preparation of people who will have an all-round development, an all-round training, people who will be able to do everything'. And it is very much more important that everyone should be something of a doctor and a pharmacist, and able to do 'running repairs' on himself or herself than that they should be able to do so to a motor car or even a bicycle.

Unfortunately our first-aid classes do not go far enough. We should not think much of a motorist who had never looked under the bonnet of his car. But we should mostly be horrified at the idea that every child should look at the inside of a man and of a woman. Yet till this is done we shall find it difficult to take a sensible and objective view about our health.

What ought everyone to know about drugs? First of all, they can be divided into three great classes: those which kill parasites, visible or microscopic, those which destroy poisons, and those which act on our bodies. Of course, some drugs have a double-barrelled action, and others, such as hypophosphites, have none at all, except the economic effect of transferring money from the patient or his approved society to the chemist and manufacturer.

The danger with germ-killing drugs is that they may kill the patient as well as the germ. For example, antiseptics such as phenol ('carbolic acid') and mercuric chloride ('corrosive sublimate') are very valuable for killing germs on the skin before an operation. And they are used in treating infected wounds. But they may be absorbed from wounds or when swallowed, and have killed a number of people.

Among the most important antiseptics for external use are those like calomel ointment and protargol, which can be used to prevent the transmission of venereal diseases from one person to another. They are outrageously expensive, and it is illegal to sell them with instructions for their use, so these diseases are commoner than they would be if the antiseptics were correctly employed.

Other drugs act on parasites inside our bodies. Worms are no great danger in England, but kill millions of people in the tropics. All the drugs which kill them, such as antimony tartrate and male fern extract, are poisonous to man in large doses, and they should never be used without medical supervision. Quinine kills off the parasite causing malaria, emetine (from ipecacuanha) that of amoebic dysentery, and salvarsan that of syphilis. These are all of them single-celled organisms which can wriggle about, and are probably best considered as animals, while most infectious diseases are caused by bacteria, which are regarded as plants, though the distinction is not very sharp.

Until recently there were very few ways of killing bacteria in the body unless they were concentrated in a small wound or abscess or in some special region such as the urinary tract. But recently an immensely powerful group of drugs, including sulphanilamide and its derivatives, have been put on the market. The best known of these has the trade name Prontosil. This has cured a number of cases of septicaemia ('blood poisoning' and puerperal fever) in the most dramatic way. It also acts on local infections, such as gonorrhoea and some kinds of abscesses. But it is a dangerous drug, and has killed a number of people.

The ideal germ-killer has not been found, and no germ-killing drugs, except quinine, should ever be taken by the mouth except under supervision, still less injected. For the dose needed to kill a man is only a few times greater than that needed to disinfect him.

PAIN-KILLERS

In the last article I wrote of the very few drugs which actually kill germs in the body. Still fewer neutralize the poisons produced by them. This can bee done by antitoxins in a few diseases, such as diphtheria; but they have to be injected, and are quite useless when taken by the mouth.

Almost all drugs are taken for their action on the human body. And the vast majority of those which are of any real value will kill you if you take five or ten times the dose which is neeeded to produce the desired effect. This is so, for example, with morphine and digitalis, which are therefore scheduled as poisons and only administered by a doctor or nurse.

Probably the medicines which are most widely taken are purgatives, whose use has become a habit with many people. One very common type are the so-called 'liver salts', 'fruit salts', and the like, which are sold with various exaggerated claims at enormous profits. They all act in the same way. They cannot be absorbed from the intestine into the blood, and are therefore got rid of. They do not 'purify the blood', as often claimed, because they are not absorbed into it.

In nine cases out of ten, the reason why these things are needed is a social reason. Many of us do not take enough regular exercise, and also eat a diet which leaves too little residue behind. So a return to a more natural life, in which every healthy adult did even an hour's hard work daily, and our diets contained plenty of vegetables and fruit, would ruin the 'health salt' merchants. Meanwhile, one can buy enough sodium phosphate or potassium bitartrate at a chemist's to do what is required for a very small fraction of what one pays for proprietary articles.

Besides these, a great variety of substances are taken for the same purpose which act by irritating the bowels. Phenolphthalein, which is sold in about half a dozen 'patent medicines', is usually fairly harmless, but occasionally causes a strange skin eruption, with purple or pink patches which will last for years. Needless to say, this fact is not advertised.

Another group of drugs is sold for 'indigestion'—that is to say, burning or gnawing pains after meals, especially meals rich in sugar. These drugs mostly consist of magnesia, which neutralizes the acid juice in the stomach which is normally there, and only irritates you if your stomach is in some way abnormal. Thus, although they relieve pain, they do not take away its cause.

The commonest cause of gastritis—that is to say, an inflamed and irritable stomach—is worry and anxiety. It is particularly common among busmen and travelling salesmen. I had it for about fifteen years until I read Lenin and other writers, who showed me what was wrong with our society and how to cure it. Since then I have needed no magnesia. But these pains may also be due to gastric ulcer or even cancer. So it is better to consult a doctor, even though he will probably recommend magnesia. But the *Daily Worker* may effect a permanent cure.

The large majority of drugs act on the nervous system. A great many of them depress its activity. Some of them, such as chloral, sulphonal, and veronal, lead to a condition not unlike natural sleep, and are given for insomnia. But in the long run they may all lead to habit formation.

The great chemical trusts which produce these drugs use a propaganda upon the medical profession which is often just as unscrupulous as that of the 'patent medicine' sellers against the general public. New drugs are constantly being turned out which differ very little from the well-known ones, but are advertised as much safer, until they in turn are found out.

Two of these sedatives can be bought at any chemist's, and are worth having in the house. One is sodium bromide, which was till recently the standard treatment for epilepsy, and is a valuable remedy for 'nerves', in the sense of jumpiness. It is also a fairly good prophylactic for seasickness. It will often relieve sleeplessness. And it has a great advantage. If you take too much you feel miserable and depressed and may come out in very nasty spots. So it very rarely forms a habit.

Paraldehyde will give sleep as effectively as chloral or veronal. And, unlike chloral, people with heart-disease can take it. But its taste and smell are so unpleasant that no one is likely to take it for fun, and it is not often a drug of addiction.

Most drugs which act on the nervous system paralyse some particular part of it. For example, when you bend your elbow you are doing two things at the same time—namely, contracting the biceps muscle of your upper arm and relaxing the triceps muscle on the other side of the bone, which pulls the joint in the opposite direction. Now, if you are poisoned with strychnine the relaxation does not occur. On the contrary, the two antagonistic muscles pull in opposite directions and you get an appalling cramp.

In just the same way the upper parts of your brain are largely engaged in inhibiting the activity of the lower parts, which are concerned with simple activities and simple emotions, such as rage. Many drugs act in the same way as alcohol, and paralyse the higher parts of the brain to a greater or less extent, so that you tend to behave less like a man and more like an animal. This may be quite a good thing if it does not go too far, as it is quite good to take an occasional holiday and live like a healthy animal for a few days.

Many of these depressants also relieve pain. Morphine and heroin are particularly useful in this way, and the latter is remarkably useful in dealing with the intractable tickle of a severe cough. Unfortunately, some people like their psychological effects. I do not. I have taken a large dose of heroin four times a day for ten days or so without getting any 'kick' out of it or losing an hour's sleep when I stopped taking it.

And I suspect that in a decent society, where no one wanted to take refuge from reality in dreams, these drugs could be sold openly as pain-killers. The Japanese have used morphine and heroin with considerable effect to demoralize the Chinese population. I very much doubt whether they would be able to make many addicts among people under forty years old in the Soviet Union, even if they were allowed to try.

So while it is quite right for the League of Nations to fight against the trade in such drugs as these, two things are worth pointing out. They are not habit-formers with everyone. And their illicit manufacture and sale brings in such huge profits that it is most unlikely to be suppressed in those parts of the world where the whole social system is based on profits.

In countries where there is an illicit drug trade, doctors are rightly cautious of using morphine as it should be used. And in consequence there is a vast amount of quite unnecessary pain. This pain is not due to the fact that we have a choice between relieving pain and ruining character. It is just one result of an immoral economic system.

HOW TO WRITE A POPULAR SCIENTIFIC ARTICLE

MOST scientific workers desire to spread a knowledge of their subject and to increase their own incomes. Both can be done by writing on science for the general public. If one can sell the article abroad, one can also be an 'invisible export'. In what follows I shall give some hints on how to do it. But let no reader suppose that my method is the only one. Literary synthesis is like organic chemical synthesis. The method to be adopted depends on the product required, the raw materials, and the apparatus available. As my brain is my apparatus, and different from yours, my methods will also differ from yours.

The first thing to remember is that your task is not easy, and will be impossible if you despise technique. For literature has its technique, like science, and unless you set yourself a fairly high standard you will get nowhere. So don't expect to succeed at your first, or even your second, attempt.

For whom are you writing? This is even more important than the choice of subject. For you will not get an article on the history of eighteenth-century physics into a daily newspaper. *The Times* is unlikely to publish a sympathetic account of Soviet work on mineralogy, nor the *Daily Worker* a highly commendatory report on cotton breeding in the British Sudan.[1] Moreover the length of your article will depend on where it is to be published.

Now for the subject-matter. You may take a particular

[1] *The Times* has since changed its policy. But unfortunately the Sudan is still some way behind Peru, the USA, the USSR, and the West Indies in cotton breeding.

piece of research work, or a particular application of science. Or you may choose some general principle, and illustrate it from different branches of scientific work. For example an excellent article could be written on fruitful accidents. Priestley broke a thermometer, and the fate of the mercury from it led him to the discovery of oxygen. Takamine spilled some ammonia into a preparation from suprarenal glands, and crystallized out adrenalin. Probably you will do better to begin on some more specialized topic, unless you are a student of the history of science.

Remember that your treatment of it must be highly selective. So far you have probably written two main types of article. Firstly answers to examination questions in which you tried to show how much you knew about some topic. And secondly scientific papers or technical reports which dealt very exhaustively with a small point. Now you have to do something quite different. You are not trying to show off; nor are you aiming at such accuracy that your readers will be able to carry out some operation. You want to interest or even excite them, but not to give them complete information.

You must therefore know a very great deal more about your subject than you put on paper. Out of this you must choose the items which will make a coherent story. A number of the articles which are submitted to me from time to time are far too like examination answers. They give the impression that the author has looked his subject up, and tried to give a condensed summary of it. Such a summary may be all very well in a textbook, but will not hold the attention of a reader of popular articles, who does not contemplate severe intellectual exertion.

This does not mean that you must write for an audience of fools. It means that you must constantly be returning from the unfamiliar facts of science to the familiar facts of everyday experience. It is good to start from a known fact, say a bomb explosion, a bird's song, or a cheese. This will enable you to illustrate some scientific principle. But here again take a familiar analogy. Compare the production of hot gas in the bomb to that of steam in a kettle, the changes which occur in the bird each year to those which take place in men once in a lifetime at puberty, the precipitation of casein by calcium salts to the formation of soap suds. If you know enough, you will be able to proceed to your goal in a series of hops rather than a single long jump.

If you try to write an article in this way, you will probably discover your own ignorance, especially of quantitative matters. How completely do a robin's gonads revert to an infantile condition in autumn? How much more calcium is there in milk than in London tap-water? What is the maximum temperature in an exploding bomb? It may take you twelve hours' reading to produce an intellectually honest article of a thousand words. In fact you will have to educate yourself as well as your public.

When you have done your article, give it to a friend, if possible a fairly ignorant one. Or put it away for six months and see if you still understand it yourself. You will probably find that some of the sentences which seemed simple when you wrote them now appear very involved. Here are some hints on combing them out. (Remember, by the way, that I am only giving my personal opinions. Prof. Hogben writes sentences longer than some of my paragraphs, and his books sell very well, as they ought to.) Can you get in a full stop instead of a comma or a semicolon? If so, get it in. It gives your reader a chance to draw his breath. Can you use an active verb instead of a passive verb or a verbal noun? If so, use it. Instead of 'It is often thought that open windows are good for health', or 'There is a widespread opinion that open windows are good for health', try 'Many people think that open windows are good for health'. Or 'Most people', if you think that is the case.

Try to make the order of the phrases in your sentence correspond with the temporal or causal order of the facts with which you deal. Instead of 'Species change because of the survival of the fittest' try 'The fittest members survive in each generation, and so a species changes'. Not that I like the phrase 'a species changes'. It would be better to say 'the average characters of the members of a species, such as weight or hair-length, change'. Of course in the history of scientific discovery an effect is commonly known before its cause. And fairly often a mathematical theorem is known to be probably true before it is formally proved. If you enunciate your theorem before you prove it you are apt to give the impression, as Euclid does, that you are producing rabbits from a hat. Whereas if you lead up to it gently you create less impression of cleverness, but your reader may find your argument much easier to follow.

In a scientific, and still more, a mathematical paper, elegance of presentation, which often means the hat-and-rabbit method, is always great fun, and sometimes desirable. How delightful to produce some wholly unexpected function at the last moment by contour integration, to damn a suggested mechanism by an appeal to Hearnshaw's theorem, or to label a plant which won't breed true as just another case of balanced lethals. By doing so you may help the serious student to short cuts in thinking. But you will merely dazzle the ordinary reader. Go slow, and show him as many steps as you can in your argument or causal chain, even if, in your own thinking, you skip some of them or take them backwards.

When you have written your article it may seem rather gaunt and forbidding, a catalogue of hard facts and abstract arguments. A critic may say it needs padding. I object to padding for padding's sake. It is characteristic of writers who are more interested in their style than their subject-matter, such as Charles Lamb or Robert Lynd, but out of place in a scientific article. On the other hand you must do what you can to help your reader to link up your article with the rest of his knowledge. You can do this by referring to familiar facts or to familiar literature. I have been severely criticized for 'dragging in' references to Marx in my articles in the *Daily Worker*, though I think I refer to Engels more frequently. But a number of my readers are familiar with the works of these authors. Engels said certain things about change, as Heraclitus said very similar things before him, and Bergson and Whitehead after him. But for one of my readers who has read Heraclitus, Bergson, or Whitehead, a hundred have read Engels, so I prefer to cite him. If I were lecturing on the same matters to classical scholars I should perhaps cite Heraclitus, even though I think Engels said it better.

In my last book on genetics,[2] there are seven quotations from Dante's *Divine Comedy*. I have been criticized for 'dragging in' Dante. But I think it worth while to show the continuity of human thought. I don't agree with Dante's theory that mutations are due to divine providence, but I consider it desirable to point out that he had a theory on this subject. I think that popular science can be of real value by emphasizing the unity of human knowledge and endeavour, at their best. This fact is hardly stressed at all in the ordinary teaching of science, and good popular science should correct this fault, both by showing how science is created by technology and creates it, and by showing the relation between scientific and other forms of thought.

A popular scientific article should, where possible, include some news. I try, as a general rule, to include one or two facts which will not be familiar to a student taking a university honours course in the subject in question, unless his teachers keep well up with the periodical literature. As there is often a lag of five years between the publication of a

[2] *New Paths in Genetics* (Allen & Unwin, 1941).

discovery and its inclusion in a textbook, this is not very difficult in peacetime. But it is not very easy at present, in view of the number of libraries which have closed down, and the absence of many European and some American periodicals. Of course some care is needed in appraising new work. A very large number of alleged discoveries are not confirmed by subsequent workers. One well-known English popularizer of science has a perfect genius for picking out discoveries of this kind for announcement to the public. If, like myself, the writer is actually engaged in research, and has seen a number of his own bright ideas go west, he is less likely to fall into this particular trap.

In the early stages of popular writing it is well to write out a summary of the article, though I rarely do so myself. Here is a possible skeleton for an article on cheese.

Introduction. A well-known fact, say the shortage of cheese.[3]

Central theme. The process of cheese manufacture.

Why it is important. Cheese as the cheapest food containing large amounts of 'good' protein. Vitamins and calcium in cheese.

Connections with other branches of science. Rennet compared with other enzyme preparations used in industry, for example in confectionery and tanning. Other uses of specific micro-organisms, for example in brewing. Why putrid cheese is safer than putrid meat.

Practical suggestions. How to increase our cheese output. Combating mastitis in cows. Cattle-feed and fertilizers. Should cargo space be devoted to cheese rather than meat? Need for scientific planning of national food supply.

How much of this you can get in depends on the length of your article and your capacity for compression. If you are writing for a highbrow journal you may quote the passages

[3] This has lessened since this article was written.

on cheese from Euripides' *Cyclops*, if for a lowbrow, any of the jokes about the smell of cheese.

That is one way of doing it. But other writers would show cheese as part of the Mysterious Universe. We do not understand protein synthesis, nor the extreme specificity of some enzyme actions. Cheese-making is part of the pre-scientific activities by which we still keep a communion with nature. Cheese is a natural food, and beef is not. And so on. I think this is an anti-scientific attitude. But you can sell that sort of stuff, and get over a certain amount of genuine knowledge while doing so. Everyone must write popular scientific articles in his own way. I have only described one way, and I do not claim that it is the only way, or even the best possible way.

WHAT 'HOT' MEANS

ONE reason why other people find it hard to understand science, and why scientists are apt to lose their tempers with other people, is that scientists either use ordinary words with a special meaning, or invent words of their own which ordinary people do not understand.

I don't think this can be avoided. The history of science shows what has constantly happened. We start with some ordinary word, such as 'hot', whose meaning we think we understand. On the breakfast table are a tablecloth, a knife, and a pot of mustard. The plain man says the knife is cold, the mustard hot, and the cloth neither hot nor cold. A physicist will say that none of them is hotter than the others.

But that does not mean that the plain man is talking nonsense. He certainly gets a feeling of cold from the knife, and a feeling of heat from the mustard if he puts it on his tongue, or rubs it into his skin. The knife and the cloth are at the same temperature, somewhat below that of one's finger. But the knife conducts heat well, so it cools the finger much more than the cloth when one touches it.

The mustard, or to be accurate, one of the chemical compounds in it, excites the same nerve fibres in my tongue as are excited by hot substances, and gives me a sensation of heat. If I rub it into my skin it makes the blood-vessels dilate, and my skin does actually get hotter in a way which a physicist could measure.

Until thermometers were invented and made fairly accurate, it was quite impossible to get any definite answer to the question which of two bodies was hotter, much less to measure temperature or heat. Even now we are apt to trust our senses unduly.

The woman who runs our household, in which I perform such humble functions as dish-drying, insists on putting food on a slate shelf rather than a wooden one in the larder, because it is colder. Actually everything put in the larder reaches the same temperature after half an hour or so. Warm things cool a little quicker on slate than on wood, and that is all the difference. If the food were like a living man or animal, and had a source of heat in it, it would be colder on slate than on wood. But if our housekeeper reads this, she will not believe the argument, and merely take it as another proof that men, and particularly professors, don't understand housekeeping.

Confusions like this arise in part because we use the same word 'heat' for a sensation, and for a form of energy which causes it. We also use the word 'hot' to mean either a body which gives us the sensation, or a body with an unusually large amount of this form of energy. Mustard is hot in the first sense, and not in the second. We should avoid these confusions if we used specially invented words such as 'caloric' and 'calorous' for 'heat' and 'hot' in their scientific senses. But when scientists use such words they are often accused of talking jargon; and such words are often taken over and used in a metaphorical sense. This is happening to the words 'allergic' and 'energy' at present.

Students of Marx find this difficulty with the word 'value'. In ordinary talk we use it in a good many different senses, and if we change its meaning in the middle of an argument we talk nonsense about economics. In *Capital* Marx first discussed some of its different meanings, and then used it with one particular meaning in the rest of the book.

The same was true of 'labour' and 'labour power'. On the other hand he (or rather his translators) used the word 'force' in a rather haphazard way, as compared with its very definite meaning in modern physics. He might have used some special word to distinguish between say productive force and electromotive force or gravitational force, but no one is likely to confuse such very different meanings of the word 'force', as they certainly confuse different meanings of 'value'.

An essential feature of the progress of science is as follows. We start with a word whose meaning we think we understand, such as iron, hot, rat, race, or intelligence, and begin to investigate the things which it designates. We always find that it changes its meaning in the course of the investigation, and sometimes we have to invent new words for the things we discover.

Iron is quite a good example. In ordinary language it is used for a variety of metals with different properties, particularly cast iron, which is hard but brittle, and wrought iron, which is softer but tougher. Chemists have found that they are all mixtures consisting largely of one metal. But the metal which chemists call iron is quite soft when pure, in fact about as soft as copper. Ordinary iron and steel are a great deal harder because they are intimate mixtures of iron and its compounds with carbon and other elements. In fact metallurgists use such words as austenite, martensite, and cementite to describe what is generally called iron or steel.

The ordinary word rat covers two different species in England, and several more in other countries. The habits of our two rats are quite different, and the first thing to find out if you want to rid a building of rats is which species is infesting it.

As for the word race, it has so many different meanings as to be useless in scientific discussion, though very useful for getting members of the same nation to hate one another. If it means a group of people with very similar inborn physical characters, especially of skin and hair, then we ought to talk of red-haired people as constituting a special race. If it means people who talk similar languages, as when the phrases Aryan race or Slavonic race are used, then American negroes and whites belong to the same race, and Basques to another.

In fact our discussions of race are still at the pre-scientific level, as would be a discussion of whether the mustard at breakfast was hotter than the tea. The same is true of most discussions of intelligence. No doubt our descendants will be able to treat these matters scientifically. But we cannot do so yet, and we should be extremely suspicious of people who say that they can. They are very often trying to do in Britain what the theoreticians of the Nazi movement did in Germany.

CATS

WE know roughly how many adult dogs there are in England, because apart from sheep-dogs and a few others, they are taxed. We do not know how many cats there are. Recently Mr Matheson, of the Natural Museum of Wales, got the help of a number of schoolchildren to count the cats in certain areas of Cardiff and Newport. Each child reported the number of adult cats in its home.

He concludes, from these and other statistics, that the number of cats in an area is roughly proportional to the number of human beings, and not even roughly to the number of acres. The number of cats with a fixed abode is about 10½ per cent of the number of humans, the number of strays about 2½ per cent. So there are probably about five million cats in Britain, apart from young kittens. However, the number of cats per hundred human beings is nearly three times higher in the slums round Cardiff docks than on the housing estates. A cat on a housing estate may be a luxury, but in some slums she is absolutely needed to keep down the mice and rats. So if we get the houses we want, the number of cats in England is likely to go down. For it is now possible to build a house so as to give mice nowhere to live, and to make cupboards mouse-proof.

I am interested in cats for a special reason. The colour and length of their hair varies a great deal. But they are all of much the same size and shape, apart from an occasional short-tailed Manx. There are no peculiar shapes like the greyhound and dachshund, no giants like the cart-horse, or dwarfs like the Shetland pony. Further, their matings are mostly governed by their own choice, not ours. So the cat population is much more like a human population, where a fair tall woman can marry a short dark man if she wants to, than it is like a population of dogs, sheep, or horses. Hence a study of inheritance in cats will be more help than a study in dogs to understanding inheritance in man.

At least one of the colour differences in cats seems to have been originally a racial difference. The wild cats of Scotland and Europe are generally tabbies. But the cats shown in ancient Egyptian paintings, and those whose mummies have been found, were almost all yellow, or as they are commonly called, ginger cats. On the other hand I know of no evidence that there is or ever has been a race of cats all of which are black, or blue, though I am told that blue cats are much commoner both in Palestine and in Brittany than in England.

Things are much the same with man. There are countries like West Africa where all the native inhabitants have short hair and dark skins, others like England where all have long hair (if it is allowed to grow) and light skins. But other quite common characters are never characteristic of a whole race. Thus no one has ever found a race all of whom had red hair and freckled faces, though it would be easy enough for a führer who was a 'man-fancier' to breed one.

We know a lot about how characters are inherited in cats, but not enough. There are two kinds of cat which I want badly. One is a tortoiseshell male (uncastrated, of course). Tortoiseshell females are common, but males are rare, and we do not know what character their children inherit, or why, as is often the case, they are sterile. The other kind is an albino, that is to say a white cat with pink, not blue or yellow, eyes. I can guess how it would breed, but I am not sure. If any readers can get me either kind, I shall be most grateful, and quite prepared to pay. But please write before sending any cats!

Why do we find it so much easier to make friends with cats and dogs than with other mammals of about the same size? The dog has a strong inborn tendency to social behaviour, learns to obey orders, and even develops something like a conscience. But the cat is not very social, and has little signs of conscience.

One reason is that cats and dogs have sensations much more like our own than those of hoofed animals such as horses, cattle, deer, and pigs, or rodents such as rabbits and rats. There are areas on the outside of the human brain concerned with sensations, not only from the eyes, ears, and so on, but from all parts of the skin.

We know this in several ways. Injury to a part of such an area does not abolish all sensation from the corresponding skin area, as when the nerves from it are cut. But it destroys its detail, so that the patient cannot say whether he is being touched at one point or several, or distinguish a penny from a matchbox by touch. And if the brain is exposed during an operation, then if the patient is conscious, stimulation of the appropriate part gives rise to sensations felt in the corresponding skin area; while stimulation of the skin causes electrical oscillations in the corresponding part of the brain, even in an anaesthetized man. In a cat, dog, or monkey, we can use this last method to determine the areas in the brain concerned with skin sensation. The animal is anaesthetized before the brain is uncovered, and killed under the anaesthetic; so it feels no pain. The results are similar to those in man.

But in many other animals Professor Adrian has found that most of the skin is not represented by sensitive areas on the rind of the brain. In a sheep, for example, only the mouth and feet are so represented. The pig has a large area for its sensitive snout; as a man does for his hands, which have as big a part of the brain at their service as the skin of the whole trunk. In consequence a sheep or pig gets very little detailed information from most of its body, while a cat or dog does. A cat likes being stroked. To please a pig you have to scratch it with a hard stick. A cat or dog can be gentle with its whole body. A horse can only be so with its sensitive muzzle. So dogs and cats can play with us, and we with them. In fact they play with children very much as equals, and quite understand that they must not use their full strength.

Some relatives of the domestic cat, such as the Scottish wild cat, certainly show no tendency to gentleness of behaviour, but others do so. The puma 'Bill' at London Zoo before the war enjoyed tearing up newspapers, but would hold one's hand in his mouth without biting it. Unfortunately he was intelligent enough to understand that trousers do not feel, so he ruined a pair of mine without hurting the leg under them. I have little doubt that pumas could be made as safe domestic animals as our large races of dog. The most hopeful of all is probably the North American skunk. This animal defends itself when attacked by making a smell which will paralyse a man or a dog. It only bites in the last extremity. So if the scent glands are removed, which is not difficult, it makes an excellent pet. There are, in fact, probably quite a number of wild species which could become as good friends of man as our cats. If they are not domesticated before the spread of agriculture wipes them out, the loss will be irreparable.

SOME REFLECTIONS ON NON-VIOLENCE

I AM a man of violence by temperament and training. My family, in the male line, can, I think, fairly be described as *Kshattriyas*. Before 1250 our history is fragmentary. From 1250 to 1750 we occupied a small fort commanding a pass leading from the hills to the plains of Scotland. Our main job was to stop the tribal peoples of the hills from raiding the cattle of the plainsmen; but perhaps once in a generation we went south to resist an English invasion, and at least two of my direct ancestors were killed while doing so. Even when Scotland had been united to England by a royal marriage and the tribals had been pacified, the tradition persisted. When I was a child my father read to me Scott's *Tales of a Grandfather*, which are legends of the warlike exploits of the Scottish nobility, and trained me in the practice of courage. He did not do so by taking me into battles, as his ancestors might have done, but by taking me into mines. I think he first took me underground when I was four years old. By the time I was about twenty I was accompanying him in the exploration of a mine which had recently exploded, and where there was danger from poisonous gases, falls of roof, and explosions. So when in 1915 I was first under enemy shell fire, one of my first thoughts was 'how my father would enjoy this'.

I find many of the virtues and vices of the heroes of Indian epics quite intelligible and even sympathetic. The second word of the *Gita*, *dharmakshetre*, gives an exact description of my feelings when I went to the trenches for the first time in 1915. I was well aware that I might die in these flat, featureless fields, and that a huge waste of human values was going on there. Nevertheless I found the experience intensely enjoyable, which most of my comrades did not. I was supported, as it were, on a great wave of *dharma*. The European *Kshattriya*, or knightly, virtues include a detestation of various kinds of meanness, and a hatred of violence against the defenceless. The European knightly vices include an addiction to gambling. I understand Yudhisthira's point of view. A *Kshattriya* should never feel secure. His *dharma* implies that he must be prepared to risk his life, and lose it if necessary, at a moment's notice. He must therefore be prepared to risk his property. I confess that I have less sympathy with his staking his wife and his brothers.

In the war of 1914–18 I was on several occasions pitted against individual enemies fighting with similar weapons, trench mortars or rifles with telescopic sights, each with a small team helping him. This was war as the great poets have sung it. I am lucky to have experienced it.

We have now to consider two facts. The *Gita*, which is an exhortation of Arjuna to violent conduct, was the favourite poem of Gandhi, the great exponent of non-violence. War has changed its character completely in my lifetime. Modern war has two principal forms. One form is characterized by the wholesale massacre of defenceless civilians with atomic bombs and other weapons. The other, which is going on in Algeria, Malaya, Kenya, Cyprus, and other regions, is characterized by the use of ambushes and individual murder by the less armed side, and the killing of prisoners and the enslavement of whole populations by the more strongly armed side.

Modern war does not evoke any of the *Kshattriya* virtues except courage. But yet these virtues are absolutely needed in modern life, as Gandhi saw. The contradiction is of course latent in Hindu mythology. Not only Rama and Krishna, but even Buddha, the great preacher of non-violence, were *Kshattriyas*. But Parasurama, the son of Jamadagni, another avatar of Vishnu, had devoted his life to the extermination of that caste. How then can we combine the *Kshattriya* virtues with non-violence? Gandhi gave one answer to this question. There are other answers, quite compatible with Gandhi's answer, but in different spheres. Gandhi was always concerned in struggles between human groups. He did his best to eliminate violence and hatred from them.

There is another kind of struggle. I quote St Paul's letter to the Ephesians, the translation from the Greek (from memory) being my own:

'For our struggle is not against blood and flesh, but against first principles, against powers, against the *lokapalas* of the *kali-yuga*, against the spiritual sources of evil in the heavens.' I translate his word *kosmokrator* or world governor, as *lokapala*. The phrase translated as *kali-yuga* means literally 'the darkness of this age'. I think that the notion of the *lokapalas* had reached Western Asia from Buddhist sources in St Paul's day.

Some of us struggle against the natural forces which in India are too often worshipped as minor deities, for example cholera and smallpox. My father was mainly concerned with such matters as the ventilation of factories and mines, which is important both in safeguarding health and preventing explosions. When he wished to investigate why men died after colliery explosions when they had received no physical injury, he first examined dead men and horses after underground explosions, convinced himself that they had died of carbon monoxide poisoning, and then proceeded to poison himself with this gas. That is to say he breathed a known amount of it until he had fallen over unconscious, and a colleague pulled him out of the gas chamber. In this way he found out how long it takes for a given amount of this gas to overcome a man. He also found that small birds are overcome much more quickly than men (and recover much more quickly). He was however averse to experiments on animals which were likely to cause them pain or fear (carbon monoxide poisoning causes neither). He preferred to work on himself or other human beings who were sufficiently interested in the work to ignore the pain or fear. His experiments on the effects of heat could perhaps be called *tapas*. He found that he could live in dry air at 300 °F. At about this temperature his hair began to singe when he moved it.

But I do not think his motivation was that of an ascetic practising *tapas*. He achieved a state in which he was pretty indifferent to pain. However, his object was not to achieve this state but to achieve knowledge which could save other men's lives. His attitude was much more like that of a good soldier who will risk his life and endure wounds in order to gain victory, than that of an ascetic who deliberately undergoes pain. The soldier does not get himself wounded deliberately, and my father did not seek pain in his work, though he greeted a pain which would have made some people writhe or groan, with laughter. I think he would have agreed with the formulation that the *atman* or *buddhi* in him was laughing at the *ahamkara*.

I have tried to imitate him. I have drunk or breathed considerable amounts of various poisons, certainly more than half the fatal dose in some cases, and have done similar experiments on other human volunteers, including my wife. For this reason I feel a certain annoyance when I am excluded from a temple of Siva, who, according to a well-known legend, drank poison to save the other gods. If Siva exists, he may be more pleased by such an action than by the recitation of a lakh of *mantras*.

I believe that this non-violent approach to experimental biology is a fruitful one. I do not condemn those who do experiments on animals which involve their death, or even moderate suffering. But I have never done an experiment on an animal of a kind which I have not previously or subsequently done on myself; and I hope I never shall. I have dissected dead animals. But I have left instructions that my own body should be dissected by medical students. And if I die in India I hope some future Indian doctors will have the unusual experience of dissecting a European.

One great advantage of working on oneself or a friend is that far greater accuracy is possible than is usual in experiments with animals. If an animal is in pain or fear, for example, its heart is likely to beat faster and its rate of breathing may also increase. This will make it impossible to measure the effect of a drug on its pulse rate or breathing with great accuracy. But when one has done a number of experiments on oneself, one can do them on animals with some confidence that the results will be as accurate as if they were done on men.

When we were in India in 1954 my wife and I did a number of experiments of this kind on three *koi* fish. When these fish are put in foul water containing little oxygen they swim up to the surface and breathe air. We teased our fish in various ways by altering the composition both of the gases dissolved in the water in which they swam and the air above it which they breathed. I don't say these fish never suffered at all. To judge from my own experience they may have had severe headaches for some minutes. But they were certainly not seriously injured, for all three of them are alive and well in London today.

I am not at all a saintly person. I have killed animals, and eat meat, though not very much. My attitude to animals is more like that of Yudhisthira. He had killed and eaten plenty of deer. But when he was asked to enter *Svarga*, leaving the dog which had been his companion on his last pilgrimage to die on the mountains, he found, perhaps to his surprise, that this was something which, as a *Kshattriya*, he could not do. No more could I or my wife have given the *koi* fish which we had watched for two months to someone else to kill and eat.

I want to urge that this kind of attitude to animals should commend itself to Indian biologists. Unfortunately it is rather rare. I think that it should be the rule. But of all the Indian biologists whom I know the man who comes nearest to it is a Muslim, your great ornithologist, Mr Salim Ali. He is prepared to admit that he has occasionally shot birds, but he greatly prefers studying them when alive.

Frankly I regard his attitude as a challenge to Hindu biologists. There is a very great opening for non-violent biological studies in India, and, what is important, they require no complicated apparatus. Let me give some examples of what could be done. Do your songbirds sing their songs if they are brought up from the egg by human beings? Or do they have to learn their song as a human being learns a language? In Europe and North America we know that some birds must learn their song, while others produce it untaught, as some mythical Hindu characters were born with a knowledge of the Vedas. If they have to learn, some species learn from their fathers. Our English robin sings very little while helping his wife to keep the eggs warm and to feed the children which emerge from them. But when the children are just learning to fly he bursts into song again for a few weeks. This may be an expression of paternal pride, but it gives his sons the chance to learn from him. Other birds do not learn till they are nearly a year old, when they learn from other males. To bring up young birds till a year old requires an aviary where they can fly about, and a great deal of devotion. The latter is common in India, but it is commoner among illiterate people than among biologists.

If ever I settle down in India I hope to continue the study of the almost non-violent branch of biology called animal genetics. If, for example, I want to breed ducks, the conditions in many parts of India, including much of West Bengal, are ideal. One would need some wire netting to prevent unwanted matings, and perhaps a balance, a tape measure, and a chart of colours, but no other apparatus. One would of course need land with small tanks, a certain amount of food, but above all, assistants who would take the greatest care of the ducks. Such work would certainly yield many facts of biological interest, and probably increase the egg production of your ducks. Genetics is not, of course, completely non-violent. We breed a lot of flies in my laboratory, of a species which normally lives on sap from injured trees. If we let the surplus ones out in London they would starve to death. We therefore anaesthetize them and drown them in oil before they recover from the anaesthetic. I sympathize with, but do not share, the Hindu practice of non-violence to insects. They will die in any case, but they need not suffer. I am ashamed if I cause an insect suffering, but not if I kill it painlessly. It would be quite possible to practise insect genetics without offending the scruples even of a Jain. But it would mean liberating animals most of which would starve to death.

Gandhi was quite clear that men have a duty of non-violence to animals. And there is no reason why biological research in India should not be conducted on Gandhian principles. On the contrary, there is every good reason why much of it should be so conducted. If Indian physiologists were ashamed to do an experiment on an animal which they could do on themselves, Indian physiology would, in my opinion, be considerably more fruitful. The most effective method of mosquito control is not to kill mosquitoes, but to give them no opportunity of laying eggs. If you have too many cows, you should try to breed cows which will go on giving milk for more than a year after a calf is born, as some of our European breeds can. You will then have fewer cows, and more grass for each cow to eat. If your traction is largely mechanized in future you will want fewer male calves. My wife has a male fish who (as predicted before he was tested) has begotten only daughters. There does not seem to be any intrinsic impossibility in producing bulls who would do the same, though many lifetimes of human research may be needed before this is achieved.

Thousands of Indians die of snake bites every year. But the cobra is one of the most beautiful of animals, and I should be very sorry to have to kill one. It is just as possible to immunize the rural population of India against the poisonous snakes which live in their neighbourhood as to vaccinate them against smallpox. But the methods for doing so safely and on a large scale have not been worked out. This is a task for non-violent Indian biologists. Two or three of them might die of snake bite before the process was fully worked out. If so they would not have died in vain. At a later date it might be possible either to breed cobras whose poison did not kill men, or men who were not harmed by the bite of a cobra. Meanwhile I should like to meet even one Indian biologist who had immunized himself (or herself) against all snake poisons, and was prepared to answer a telephone call and remove an unwanted cobra or krait from a house without killing it.

Such ideas no doubt seem silly to many Europeans. If they seem silly to Hindus, this means that there is something badly wrong with modern Hinduism. In my opinion several things are badly wrong with it. In particular the love for all animals which is expressed in your scriptures and your art has been replaced by a series of formal prohibitions. Gandhi realized that if non-violence to human beings is to be effective, it requires both courage and intelligence. He had plenty of both. I have tried to show that courage and intelligence are needed if non-violence to animals is to be something positive, based in all cases on the love of men for animals, and in many, on the love of animals for men.

Politicians may not like this article. They may say that it calls for a diversion of effort from important to unimportant fields. I do not agree. In my experience kindness to human beings and to animals usually go together. Those who ignore suffering in animals find it easier to ignore human suffering, and conversely. I think that Indians who love animals are often perplexed because they do not see how to give practical effect to their love. I have made some suggestions, and venture to hope that some of them will be accepted. India has made many contributions to world culture. Perhaps the greatest is the ideal of non-violence. Europe's greatest contribution is the scientific method. If these can be married, their offspring may raise mankind to a new level.

APPENDIX: ADUMBRATIONS

SCATTERED through Haldane's writings—both the popular articles and scientific papers—there are hints of things to come. He sketches some experiment, or theory, or field of investigation which later and in other hands has become important. There are many possible reasons why he failed to follow up these hints himself: he was too impatient to be good at raising grants to support experimental work, the techniques needed to test an idea were not yet developed, or he simply had too many other things to think about. I give three examples, each followed by an explanation and an account of what has happened since.

HOW A GENE REPRODUCES ITSELF

From *New Paths in Genetics* (London, Allen & Unwin, 1941), pp. 43–5:

> Let us consider one of the central problems of genetics, namely, how a gene reproduces itself.
>
> When a cell divides, it produces two cells, in each of which, apart from mutation, every gene in the original cell is replaced by a similar gene. Doubtless structures outside the nucleus are also reproduced. But the method of their reproduction is different. For if structures outside the nucleus are artificially altered, this alteration is not copied. On the other hand, alterations in the genes produced by X-rays or otherwise are copied, at least in many cases. The biologist will be inclined to say that the

gene is an elementary organism, and divides to give two 'heirs' like itself. But we cannot imagine the gene swelling till it divides like an overgrown drop of water. For it does not consist of a number of like parts. If it did it could not be completely changed by the hit of a single electron. Further, the gene is within the range of size of protein molecules, and may be a nucleo-protein molecule like a virus. If so, the chemists will say, we must conceive reproduction as follows. The gene is spread out in a flat layer, and acts as a model, another gene forming on top of it from pre-existing material such as amino-acids. This is a process similar to crystallization or the growth of a cellulose wall.

Now suppose that the biologist and the chemist go round to a physicist, and ask him whether he thinks the genes in the two 'daughter' cells are the heirs of the original gene, or that one is the model and the other the copy. The physicist will have to say something like this: 'Your alternative is a false one. I can't yet put the true answer unambiguously in words, but I can put it in symbols. Here is the difficulty. How can one distinguish between model and copy? Perhaps you could use heavy nitrogen atoms in the food supplied to your cell, hoping that the "copy" genes would contain it while the models did not. But unfortunately all proteins in a living cell seem to exchange nitrogen with the fluid around them. So the most you could do would be to say that there was a certain probability of one gene being the model and the other the copy. No doubt if the cell divides quickly enough this probability is pretty high. But one can never say that either of your alternatives is completely corrrect. Remember that it is not just a question of human ignorance. On the contrary, the impossibility of distinguishing between two things is only our human expression for a relation between them which also

> manifests itself in a term in their joint energy, or, if you like that word, in a force attracting them. No doubt this attraction is very small in the case of genes. But it must be there, and it may yet prove to be important in biology, as similar attractions are in physics. So you are both right and both wrong.'
>
> I think that throughout genetics an attempt to impose mechanistic interpretations such as the model and copy theory will break down in some such way as this. However, a refutation of mechanism is not a refutation of materialism. On the contrary, even if we reject Morgan's mechanism, we must be grateful to him for showing that the gene, the physical basis of heredity, is a material object.

This passage was recently pointed out by H. F. Judson in *The Eighth Day of Creation*; my attention was drawn to it by Jonathan Howard. It is an extraordinary prevision of one of the most famous experiments in the history of molecular genetics, reported by Meselson and Stahl in 1958. In 1941, Haldane supposed that genes were proteins. By 1958 it was known that they are made of DNA, and that DNA is a double-stranded structure in which the genetic message is carried twice: in a positive form on one strand, and in a complementary, negative form as the other. It is as if I remembered my wife's appearance by carrying with me a positive transparency and a negative. By labelling with heavy nitrogen, just as Haldane suggests, Meselson and Stahl showed that, when a gene reproduces, the positive forms a new negative and the negative a new positive, so that each new double-stranded gene consists of one new strand and one old one.

There are two remarkable features of Haldane's suggestion. The first is that, with no knowledge of the double-stranded nature of genes to help him, he was able to foresee so accurately how the nature of gene replication would be elucidated. The other is that his suggestion is made in the course of a discussion of how mechanistic interpretations in biology may break down. Simultaneously, he has a clearer mechanistic picture of the gene than any of his contemporaries, and he suspects that it will fail. As it happens, the turnover of material in DNA is so slow (if it happens at all) that the difficulties Haldane envisaged do not arise: at present, mechanistic views are triumphant in molecular biology. I am tempted to say that he was misled by his Marxist philosophy, and would have done better to stick by his first, mechanistic insight. But he may yet have the last laugh.

THE EVOLUTION OF ALTRUISM

From *Penguin New Biology* 15 (1953), 44:

> . . . it is only in . . . small populations that natural selection would favour the spread of genes making for certain kinds of altruistic behaviour. Let us suppose that you carry a rare gene which affects your behaviour so that you jump into a flooded river and save a child, but you have one chance in ten of being drowned, while I do not possess the gene, and stand on the bank and watch the child drown. If the child is your own child or your brother or sister, there is an even chance that the child will also have this gene, so five such genes will be saved in children for one lost in an adult. If you save a grandchild or nephew the advantage is only two and a half to one. If you only save a first cousin, the effect is very slight. If you try to save your first cousin once removed the population is more likely to lose this valuable gene than to gain it. But on the two occasions when I have pulled possibly

> drowning people out of the water (at an infinitesimal risk to myself) I had no time to make such calculations. Palaeolithic men did not make them. It is clear that genes making for conduct of this kind would only have a chance of spreading in rather small populations where most of the children were fairly near relatives of the man who risked his life. It is not easy to see how, except in small populations, such genes could have been established. Of course the conditions are even better in a community such as a beehive or ants' nest, whose members are all literally brothers and sisters.

This passage contains the seeds of an idea which was developed by Hamilton (published in 1963, and more fully a year later), and which has since become the basis of an extensive theory of the evolution of social behaviour. It is interesting that R. A. Fisher had the same idea, but also failed to follow it up. In essence, the idea is as follows. If a gene *A*, in an individual *X*, makes it more likely that *X* will perform some act, the effect of which is to reduce *X*'s chances of survival and reproduction but to increase the chances of *X*'s relatives, then more copies of *A* may be transmitted to future generations than would be the case if the act had not been performed. This happens because identical copies of the gene *A* may be present in *X*'s relatives. All complex animal societies involve some degree of co-operation between relatives. It is now thought that this is not an accident: co-operation evolves in part because the interacting individuals are relatives.

It is clear that Haldane had grasped the essential arithmetic of the situation. The break-even point, at which the expected number of copies of a gene in Haldane is the same whether he jumps into the river or not, occurs if he saves one cousin and has one chance in eight of being drowned. I take it that he wrote of one chance in ten, because he wanted a number which made it just worth while to jump in. However, the fact that he refers to a rare gene suggests that he had not appreciated that the answer is the same (one-half for a brother, one-eighth for a cousin) whether or not the gene is rare. He saw the relevance of the idea to the evolution of social insects, but he missed the significance of the peculiar genetic system of ants, bees, and wasps, in which males develop from unfertilized eggs and have only a single set of chromosomes. It is a consequence of this that sisters have a particularly high proportion of their genes in common (in fact, three-quarters). When I first came across this point in Hamilton's 1964 paper, I felt furious with myself for not having seen it, but slightly comforted that Haldane had missed it too.

DISEASE AND EVOLUTION

From a paper in *La Ricerca Scientifica* (1949):

> It is generally believed by biologists that natural selection has played an important part in evolution. When however an attempt is made to show how natural selection acts, the structure or function considered is almost always one concerned either with protection against natural 'forces' such as cold or against predators, or one which helps the organism to obtain food or mates. I want to suggest that the struggle against disease, and particularly infectious disease, has been a very important evolutionary agent, and that some of its results have been rather unlike those of the struggle against natural forces, hunger, and predators, or with members of the same species.
>
> Under the heading infectious disease I shall include, when considering animals, all attacks by smaller

organisms, including bacteria, viruses, fungi, protozoa, and metazoan parasites. In the case of plants it is not so clear whether we should regard aphids or caterpillars as a disease. Similarly there is every gradation between diseases due to a deficiency of some particular food constituent and general starvation.

The first question which we should ask is this. How important is disease as a killing agent in nature? On the one hand what fraction of members of a species die of disease before reaching maturity? On the other, how far does disease reduce the fertility of those members which reach maturity? Clearly the answer will be very different in different cases. A marine species producing millions of small eggs with planktonic larvae will mainly be eaten by predators. One which is protected against predators will lose a larger proportion from disease.

There is however a general fact which shows how important infectious disease must be. In every species at least one of the factors which kills it or lowers its fertility must increase in efficiency as the species becomes denser. Otherwise the species, if it increased at all, would increase without limit. A predator cannot in general be such a factor, since predators are usually larger than their prey, and breed more slowly. Thus if the numbers of mice increase, those of their large enemies, such as owls, will increase more slowly. Of course the density-dependent check may be lack of food or space. Lack of space is certainly effective on dominant species such as forest trees or animals like *Mytilus*. Competition for food by the same species is a limiting factor in a few phytophagous animals such as defoliating caterpillars, and in very stenophagous animals such as many parasitoids. I believe however that the density-dependent limiting factor is more often a parasite whose incidence is disproportionately raised by overcrowding. . . .

Probably a very small biochemical change will give a host species a substantial degree of resistance to a highly adapted micro-organism. This has an important evolutionary effect. It means that it is an advantage to the individual to possess a rare biochemical phenotype. For just because of its rarity it will be resistant to diseases which attack the majority of its fellows. And it means that it is an advantage to a species to be biochemically diverse, and even to be mutable as regards genes concerned in disease resistance. For the biochemically diverse species will contain at least some members capable of resisting any particular pestilence. And the biochemically mutable species will not remain in a condition where it is resistant to all the diseases so far encountered, but an easy prey to the next one. A beautiful example of the danger of homogeneity is the case of the cultivated banana clone 'Gros Michel', which is well adapted for export and has been widely planted in the West Indies. However it is susceptible to a root infection by the fungus *Fusarium cubense* to which many varieties are immune, and its exclusive cultivation in many areas has therefore had serious economic effects. . . .

I wish to suggest that the selection of rare biochemical genotypes has been an important agent not only in keeping species variable, but also in speciation. We know, from the example of the *Rh* locus in man, that biochemical differentiation of this type may lower the effective fertility of matings between different genotypes in mammals. Wherever a father can induce immunity reactions in a mother the same is likely to be the case. If I am right, under the pressure of disease, every species will pursue a more or less random path of biochemical evolution. Antigens originally universal will disappear because a pathogen had become adapted to hosts carrying them, and be replaced by a new set, not intrinsically more

> valuable, but favouring resistance to that particular pathogen. Once a pair of races is geographically separated they will be exposed to different pathogens. Such races will tend to diverge antigenically, and some of this divergence may lower the fertility of crosses. It is very striking that Irwin (1947) finds that related, and still crossable, species of *Columba*, *Streptopelia*, and allied genera differ in respect of large numbers of antigens. I am quite aware that random mutation would ultimately have the same effect. But once we have a mechanism which gives a mutant gene as such an advantage, even if it be only an advantage of one per thousand, the process will be enormously accelerated, particularly in large populations.

This example differs from the other two, in that Haldane devoted a whole article to the topic, rather than making a brief reference in passing. Perhaps he took the matter as far as he could at that time. The ideas discussed have become increasingly important since, for both practical and theoretical reasons. Pests and disease-causing organisms evolve resistance to the chemical agents we use to destroy them. They also evolve the capacity to attack any variety of agricultural plant that is widely grown. Consequently, we are observing an arms race between the breeders of disease-resistant varieties, and micro-organisms that evolve new ways of attacking them.

The idea that diseases are responsible for much of the observed biochemical and genetic variability of wild populations is widely accepted, as is the idea that an arms race between hosts and their diseases leads to continuing evolutionary change in both kinds of organism. This has led to a further speculation, which Haldane does not mention. This is that disease may be an important reason why organisms reproduce sexually. There is little point in reproducing sexually in an unchanging world, but it may be crucial in a species which is continually facing new challenges from its parasites.

INDEX